SKEPTOID

CRITICAL ANALYSIS OF POP PHENOMENA

BY BRIAN DUNNING

FOREWORD BY JAMES RANDI

Ignorance is preferable to error; and he is less remote from the truth who believes nothing, than he who believes what is wrong.

Thomas Jefferson

Contents

FOREWORD

by James Randi

If you don't already have a skeptic's view of the world, this book will perhaps shock and dismay you. In fact, I think there will even be some skeptics who pick up this book and find it perhaps too challenging – though I hope not.

I've always said that there are two kinds of faith: blind faith requires no evidence to support it, and it is the basis of all religious conviction and what I refer to as "woo-woo" matters. Then there's faith based on evidence that provides a framework and a foundation of support. That's the kind – the only kind – that I choose to accept. The author of this book would, I'm sure, agree with me.

I also share with author Brian Dunning a profound atheism and a fearless approach to examining, criticizing, and – if need be – condemning ill-founded support for the inane, juvenile, and baseless acceptance of nonsense that we find around us in spite of the obvious fact that the scientific method has brought us what should be a welcome new understanding of so many of the former mysteries upon which our ancestors gave up so long ago. This understanding is our strength and our future, if we will accept it for what it is: an admittedly imperfect picture of the universe around us, but one which is improving every day simply because we continue to wonder about it and to improve our knowledge of it.

I have repeatedly found that persons who become relieved of their superstitions and misconceptions of the world react in a matter which is quite different from what might be expected, once they have opened the door and stepped through into a rational, thinking, world. Most of these folks have expressed to me a profound sense of relief and renewed strength, rather than a disappointment and/or a sudden fear of the new and uncaring universe they have come upon or been led to confront. Such an epiphany, in my opinion, is a gift that a rational thinker such as Brian can confer upon his readers.

I don't think he missed many subjects. Everything is here, from homeopathy to chiropractic, the Amityville Horror to the Bible Code, and "orbs" to the Marfa lights. He obviously has many well-thumbed volumes of claptrap at his disposal – and disposal might be a good way of dealing with such nonsense. The library at the James Randi Educational Foundation consists – presently – of 2,089 books of which only about 5% are rational. Those are by such authors as Sagan, Feynman, Shermer, Asimov, Gardner, and a handful of others, and I assure you that they hold their own against the others – as Mr. Dunning has obviously discovered.

Yes, you will see in this book the ill-concealed dismay and anger of the author and that is exactly what we might expect. Rationalists can easily become impatient with the sloppy approach that the public has when confronted with the plethora of nonsense that arrives daily via television, bookstores, and every conceivable media outlet that can profit from its distribution. Well, let us hope that this book receives the recognition and the distribution that it so well deserves. To recommend this book to a friend, to a school, or to a local library, is a distinct service that you – the reader – may choose to perform.

I hope so, and I encourage you to follow that suggestion.

James Randi.

INTRODUCTION: DRAGONS

In ancient times, unexplored regions on maps would often be given fearsome legends like "Here Be Dragons". Unknowns were frightening, and it gave some comfort to at least be able to label the unknown. Hypothesized dragons were a good enough explanation for what would otherwise be ungraspable. With a made-up concept and a few words, the unknowable suddenly became simple and satisfying.

Those ancient cartographers would have felt quite at home today. De facto practice among most people is still to give satisfying labels to quantify and conveniently package the unknown. When faced with a phenomenon for which one does not personally know a rational explanation, like dreaming of your uncle the night before he dies, it's much easier to accept a simple explanation like "psychic connection" than to grasp the complexities of cognitive phenomena, confirmation bias, and law of large numbers. "Here Be Dragons" is so much easier. The vast majority of the population accepts dragons – or their logical equivalents – as natural components of our world that should be taken for granted.

For the most part, any given instance of this is not especially harmful. Your friend takes a homeopathic remedy for chronic headaches, and this doesn't hurt anything (except your friend's wallet). Your grandmother asks for a blessing on her poodle that he might overcome his mange, and this does no harm. But consider it from a new perspective: multiply these by a million or a billion. Imagine widespread faith that non-evidence based modalities such as homeopathy or blessings are good enough substitutes for modern medicine. Imagine how much of the world's finite resources of time, money, funding, and attention are diverted from scientifically developed and clinically tested methods. Imagine what happens when a population, faced with some new crisis, believes that untestable alternatives that lack empirical foundation are the best foot forward. If we discovered a planet killing meteor that would strike Earth, should we use scientific methods to redirect it, or should we use that time and funding instead to all join hands and concentrate our psychic energy? This may seem like an extreme example, but it accurately represents diversions of human ingenuity that happen daily, all over the globe.

In 2006, I began the Skeptoid podcast (http:// skeptoid.com) in the hope of helping people to re-examine pseudoscientific beliefs in favor of scientific reality. The less people are willing to accept pseudoscience at face value, and the more willing they are to learn the realities of our world, the better for everyone. This book is based upon the first fifty Skeptoid podcast episodes. As its name suggests, Skeptoid consists of short skeptical factoids.

You will not find these chapters to be footnoted or otherwise supplemented with authoritative references, something for which I'm often criticized. There are two main reasons I omit footnotes. First, if you are truly interested in a subject, I want you to research it for yourself. I don't want to tell you what sources I personally found compelling. You need to do your own work; you shouldn't be believing me anyway. Second, many "authoritative" sources, especially those found on the Internet, come from someone with a particular agenda. It's easy to find a reference to support any point you want to make, and this renders every footnote in the world suspect at face value. Never simply trust that a little number following a statement[42] means that it's true.

There may indeed be undiscovered dragons in our world. But there is also something we know for a fact: We haven't found any dragons yet. We've looked in a lot of places, and seen some extraordinary things; but never yet has science been forced to throw in the towel and admit the reality of magic.

1. NEW AGE ENERGY

I'm feeling a little low today, so let's tap into a source of energy from a neighboring dimension as a quick upper.

Faith in pseudoscience is rampant. Everywhere you turn, intelligent people fully accept the existence of anything from psychic phenomena, to angels, to new age healing techniques, to ancient health schemes based on mysterious energy fields not understood by science. Most of these paranormal phenomena rely on "energy," and when the performers are asked to explain, they'll gladly lecture about the body's energy fields, the universe's energy fields, Chi, Prana, Orgone, negative energy, positive energy, and just about anything else that needs a familiar sounding word to explain and justify it. Clearly, there are too many loose interpretations of the word energy, to the point where most people probably have no idea exactly what energy really is.

I believe that if more people had a clear understanding of energy — and it's not complicated — there would be less susceptibility to pseudoscience, and more attention paid to actual technologies and methods that are truly constructive and useful.

A friend told me of her ability to perform minor healings, and her best explanation was that she drew energy from another dimension. She had recently rented *What the Bleep Do We Know,* so she was well prepared to explain that alternate dimensions and realities should be taken for granted, since science doesn't really know anything, and thus those things cannot be disproven. That's fine, I'll concede that she can make contact with another dimension: after all, the latest M theories posit that there are probably ten or eleven of them floating around, and I'll just hope that my friend's is not one of those that are collapsed into impossibly small spaces. What I was really interested in was the nature of this vaguely defined energy that she could contact.

I asked what type of energy is it, and how is it stored? Is it heat? Is it a spinning flywheel? Is it an explosive compound? Is it food? These are examples of actual ways that energy can be stored.

In popular New Age culture, "energy" has somehow become a noun unto itself. "Energy" is considered to be literally like a glowing, hovering, shimmering cloud, from which adepts can draw power, and feel rejuvenated. Imagine a vaporous creature from the original *Star Trek* series, and you'll have a good idea of what New Agers think energy is.

In fact, energy is not really a noun at all. Energy is a measurement of something's ability to perform work. Given this context, when spiritualists talk about your body's energy fields, they're really saying nothing that's even remotely meaningful. Yet this kind of talk has become so pervasive in our society that the vast majority of Americans accept that energy exists as a self-contained force, floating around in glowing clouds, and can be commanded by spiritualist adepts to do just about anything.

There is well known authority for the simple, concrete, scientific definition of energy. Take Einstein's equation, $E=mc^2$, that every schoolchild knows but so few spend the 30 seconds it takes to understand. Energy equals mass times the speed of light squared. Simplify it. Mass can be expressed in grams, and speed can be expressed in meters per second. Thus, an object's energy equals the amount of work it takes to move a few grams a few meters in a few seconds. Energy is a measurement of work. If I lift a rock, I'm inputting enough potential energy to dent the surface of the table one centimeter when I drop it. The calories of chemical potential energy that my bloodstream absorbs when I eat a Power Bar charge up my muscles enough to dig two hundred pounds of dirt in my garden. Nowhere did Einstein discuss hovering glowing clouds, or fields of mystical power generated by human spirits.

When spiritualists discuss energy, don't blindly accept what they're saying simply because energy is a word you're familiar with, and that sounds scientific. In many cases, their usage of the word is meaningless. When you hear the word "energy" casually used to explain a mystical force or capability, require clarification. Require that the energy be defined. Is it heat? Is it a spinning flywheel?

Here's a good test. When you hear the word "energy" used in a spiritual or paranormal sense, substitute the phrase "measurable work capability." Does the usage still make sense? Are you actually being given any information that supports the claim being made? Remember, energy itself is not the thing being measured: energy is the measurement of work performed or of potential.

Take the following claim of Kundalini Yoga as an example: *"The release and ascent of the dormant spiritual energy enables the aspirant to transcend the effects of the elements and achieve consciousness."* This would be a great thing if energy was indeed that shimmering cloud that can go wherever it's needed and perform miracles. But it's not, so in this case, we substitute the phrase "measurable work capability" and find that the sentence is not attempting to measure or quantify anything other than the word "energy" itself. We have a "dormant spiritual measurable work capability," and no further information. That's pretty vague, isn't it? For this claim to have any merit, they must at least describe how this energy is being stored or manifested. Is it potential energy stored in the chemistry of fat cells? Is it heat that can spread through the body? Is it a measurable amount of electromagnetism, and if so, where's the magnet? In any event, it must be measurable and precisely quantifiable, or it can't be called energy, by definition.

There's a good reason why you don't hear medical doctors or pharmacists talking about energy fields: it's meaningless. I think it's generally good policy to remain open minded and be ready to hear claims that involve energy, but approach them skeptically, and scientifically. The next time you hear such a claim, substitute the phrase "measurable work capability" and you'll be well equipped to separate the silly from the solid.

2. RELIGION AS A MORAL CENTER

In this chapter we pull open the drawer in the motel bureau and face the need to have a Moral Center, that core set of behaviors and ethics that governs the way we conduct ourselves and live our lives.

It may shock you to learn this, but I have no religious convictions. I do not believe that supernatural deities exist. There's nothing evil or wrong about that. I view the Christian God in the same way that the average Christian views Shiva, Athena, or Thetans. There's nothing evil or wrong about doubting the actual divinity of those characters either. Yet a common generalization made by some religious people is that atheists lack a moral center. More than once, in late night bull sessions with religious friends, I've been told that faith is a necessary component for developing a sound moral center. The implication is that religious beliefs play an important role in the development of a normal, healthy system of ethics and personal conduct. Without religious faith, one is less likely to become a "moral" person. Thus, one of many reasons that people of religious conviction want to reach out to atheists is to help them to find a Moral Center, so we don't have a bunch of naked godless pagans running around wreaking havoc and mayhem.

My response to the religious people — after thanking them for the assumption that I am an unethical person — is to compare our Moral Centers and see where these supposed differences lie. If you knew me personally, you would probably find me to be a generally upstanding person, like yourself, who stays out of trouble, brushes his teeth, walks his kids to school, and tries not to shout too much in the library.

Like you, I am generally an honest person. I don't cheat people in business. I don't steal or commit crimes any worse than speeding on the freeway. I lie all the time, but only when the lie is a helpful one: "Yes, you look great in those parachute pants."

SKEPTOID

Like you, I play fair in sports, even against unfair opponents. I try to be a gracious loser, and occasionally even a gracious winner.

Like you, my family is the most important thing in my life. Preserving the love, trust, and happiness in my family absolutely outweighs all other priorities in my life.

Like you, I have a clear sense of right and wrong. Generally, behavior that injures someone else is wrong, and most of us avoid doing that whenever possible.

Like you, if I see a complete stranger drop their wallet — even if they're a different race and speak a different language — I'll spring into action like Batman to return it to them. It would never occur to either you or I to keep it or expect a reward for returning it.

If I see an elderly woman, I don't run over, punch her in the face and steal her purse; and neither does a religious person. But note that no religious person ever says "I would love to punch out that old woman, but I can't because God told me not to." Nobody is going to do something like that, because it's so obviously wrong. Rarely or never does a basically good person — and that's most of us — need religious commandments to stop them from doing something wrong.

In summary, my Moral Center is essentially the same as yours. It comes from the basic goodness of human nature, and my own sense of right and wrong that is universally shared among all people. It does not stem from having read any particular set of religious commandments, or from fear of punishment from a deity. Since I formed this ethical system in the lack of a religious context, how could my Moral Center be so similar to that of the average Christian or Buddhist? I argue that everyone's basic Moral Center comes from human nature, the nurture of societal interaction, and the sense of right and wrong. Since everyone already has these things, the need to credit religion as an additional source is redundant and thus wholly unnecessary.

A common retort from religious people is that God gave me those things: common sense, and the ability to tell right from wrong. If that's so, and everyone (atheists included) has been gifted with all the fundamentals needed to develop a Moral Center, then we're still left at the same place. A religious upbringing is still superfluous.

Religion is an important and favored part of life for most people. Its practice brings them satisfaction in many ways. But religion is absolutely not necessary to become a good person, or to have a sound Moral Center. Philanthropists, educators, doctors, emergency workers, and Nobel laureates have the same general breakdown of religious affiliation (including atheism) as the population at large, because they are the population at large.

3. Rods: Flying Absurdities

From the cryptozoology files, we're going to look now at rods, those magical, mystical living UFO's that inhabit the invisible shadowlands of Earth.

Rods are said to be flying creatures, from a few inches to a few feet in length, that are invisible to humans, but visible to cameras, both film and digital, both still and video. Their bodies are shaped like long thin rods, and their only appendages are wavy wings, one on each side, stretching the full length of their bodies. They move through the air by undulating these wings, like eels swimming through water.

A gentleman named Jose Escamilla claims to be the discoverer of rods. On his web site, Roswellrods.com, he says that he first captured rods on video in 1994. He says he was taping UFO's when he accidentally filmed the rods as well. Since Mr. Escamilla did not recall seeing any such thing in person while he was taping, he decided the most likely explanation for his video is that he'd discovered a new species of flying creature that is invisible to humans, and only shows up on film or video.

Since then, innumerable photographs and videos have surfaced that purport to show rods. Search the Internet, and you'll find hundreds of them.

If rods are as ubiquitous as it would seem they are, why is their existence not generally accepted? Justification for the existence of rods requires that four basic claims be proven or at least shown to be reasonable:

❖ There should be zoological precedences for the existence of undiscovered insects up to a meter in length. New species are being discovered all the time, so I think we should grant this one. It's certainly possible that there are undiscovered flying creatures a meter in length.

❖ We must accept the existence of creatures that are invisible, although they're up to a meter in length and perhaps up to several inches wide. Discounting microscopic organisms, the natural world offers no better than transparency, such as that found in some species of jellyfish. Transparency is not invisibility. Supporters of rods have not proven that invisibility in the animal kingdom is possible, and they will need to do so by presenting an invisible animal.

❖ Certain images must be visible only in the output of all types of visible wavelength cameras, but not visible to the naked eye. When cameras output their images to the final medium, be it film, paper, or a video screen, we see their output because our eyes see the same visible wavelengths that were recorded and output. We're not talking about thermographic or other non-visible-wavelength camera technologies here, so rod supporters will need to prove that all standard cameras can convert certain invisible wavelengths into visible ones, without affecting the visible wavelengths; which is something those cameras were not designed to do. Only with this proof can it be reasonably accepted that it's possible for a camera to see something that was invisible to the photographer.

❖ Even if all of the above can be substantiated, there needs to be a lack of a more likely explanation. If a simple procedure can be shown to easily reproduce the appearance of rods on camera, then we haven't even established that there is a phenomenon to be investigated.

As you might expect, there is indeed an alternate explanation, and a simple procedure to take a picture showing rods. Picture yourself standing with the sun at your back, facing a large shaded area, such as the shaded entrance to a cave. Dragonflies (or other insects) are flying everywhere, darting back and forth at around 20mph, which is about 29 feet per second (dragonflies can hit 60mph). Take a photograph, with a common shutter speed of 1/30th of a second. In that time, the dragonfly will travel about 12 inches. Because your exposure is set for the dark background, the path traced by the dragonfly's transit will be overexposed and will appear solid white. The dragonfly will make one full wingbeat in in that time (some insects would beat their wings twenty times in 1/30th of a second), so the path described by its wingtip on your film image would be one full sine wave period, twelve inches long. There would be one of these sine waves down each side of the twelve-inch-long rod shaped track traced by the dragonfly's moving body.

This phenomenon is so common that most any professional photographer can tell you about being plagued by it while trying to take outdoor photographs or video in similar lighting conditions. Nevertheless, the resulting image is strange enough that someone not familiar with photography basics might conclude that the subject in the photograph was in fact twelve inches long with undulating wings, and the photographer would be absolutely correct in stating that he did not see any twelve inch long flying creatures with his naked eye.

The conclusion from all this is that rods are a well known, well established, and well understood byproduct of photography. The proposed alternate explanation, that they are an unknown and invisible lifeform only seen by cameras, requires that some pretty outrageous claims about invisibility and photography be proven. Until they are, or until a rod is captured and can be studied, I see no reason to suspect that such things might exist.

4. ETHICS OF PEDDLING THE PARANORMAL

This is where I break ranks with the majority of the skeptical community, and come out, surprisingly, generally in favor of those who peddle the paranormal, in cases where no harm is done.

In our society, people have the right to purchase things they want that are of no benefit, or even harmful. Cigarettes, alcohol, expensive cosmetic products containing questionable ingredients like "extract of oleander" — these are just a few examples. It's a free country, and most people want these things. We've decided, as a nation, that the preferences of a few should not curtail the freedoms of the many. And I believe most skeptics would agree: paranormal services from palm readers to homeopathy stores have every right to exist. I hope my kids don't become customers, but I feel education is a better way to address it than government intervention.

Since we agree that these services have the right to exist, and that people must be free to make their own choices about using them, I personally would have no problem stepping up and selling my own psychic predictions. I would love to be able to perform a good cold reading. My dream is to start a church and become fabulously wealthy, with the world's happiest customers. These customers are people who are already believers, whose minds are not about to be changed by a few skeptics. They are going to buy these services: and if they don't buy them from me, they're going to buy them from the psychic next door. I could do a good job. I could be perfectly convincing and tell them exactly what they hope to hear for their money. In fact, the customer's experience will be identical to that they'd receive from the "real" psychic next door. We agree that customers have the right to spend their money on whatever they want. We agree that a customer is being deceived whenever he buys any supernatural product, no matter who sells it. We agree that no power on earth could convince that customer that he's being deceived. Add it all up, and we have a customer who insists on being deceived, and who has the right to purchase that deception. I believe that it's perfectly acceptable — and perfectly ethical — for me, even as a skeptic, to take advantage and sell the same product.

If you're like most people, you're disagreeing with me. You're probably saying that I'm being dishonest and lying to the customer, while the real psychic (though his powers are no more real than mine) is at least being honest. He's wrong, but he's honest. We're selling the same thing, and both giving the customer a satisfying experience. I see it just like a supermarket manager who allows cigarettes to be sold in his store. He knows they're a bad product, but people want them, and that's the way it is. Yet I never hear my detractors criticize the supermarket manager.

The best argument I've heard against my position is that I'm taking away the customer's dignity, in removing his right to make a choice. I'm being disingenuous, telling him that I'm someone I'm not, when my psychic competitor next door is being honest in claiming psychic powers. The customer chooses to go to a psychic. I'm lying to him, while the psychic next door is not. I understand this argument, and I agree that it's true. But the reason this argument doesn't convince me is that it's irrelevant — the net result is exactly the same. My personal beliefs have no bearing on the transaction (just like the supermarket manager), and focusing on this question is ignoring the elephant in the room: the person wants to buy nonsense. The personal feelings or opinions of the person selling it are simply not part of the equation.

Now, it's time to address the point that's probably foremost on your mind. What about the cases where the pseudoscience being purchased is either harmful, or takes the place of essential medical or psychiatric care? I said at the very beginning: I'm generally in favor of those who peddle the paranormal, in cases where no harm is done. And this is the vast majority of cases. What about the exceptions?

Here's a hypothetical case where the customer really needs medical care: they have treatable cancer, but prefer to pay me for New Age healing by the laying on of hands. I assure you that I am neither completely stupid, nor irresponsible, nor in any particular need of blood money. In this case, I would put on my best New Age hat, and explain to this person in New Age terms that I hope they would understand and accept, that New Age healing can only help when applied alongside conventional cancer treatment. I'm smart enough to realize that if I tell him New Age healing is bunk and he should go to the doctor, he'll write me off as a debunker and not listen, and go instead to the psychic next door. Here is where my New Age services are better — infinitely better — than those of the "real" psychic, who genuinely believes that laying on of hands should be used to the exclusion of real medicine. And people tell me that I'm the one being unethical. The "real" psychic in this case should be imprisoned.

It's the same in cases where the customer needs psychiatric care. Let's say his mother died, and for some reason he has developed real psychological problems, and wants me to contact his dead mother. This is not someone who wants me to predict tomorrow's horse race, this is someone who probably needs help beyond my pretended abilities. In this case, I'd dim the lights, hold as convincing a seance as I could, and tell him that his mother is worried about him and begs him to seek some professional help. If you tell him in this manner, he's likely to actually listen, and the doctor can handle it from there. If you take the usual skeptical path, and explain to him that talking to the dead is bunk and only a real doctor can help him, he won't listen, he'll go to the "real" psychic next door, and his problems will continue. Again, my services are good because they'll actually lead to a professional solution; the "real" psychic's services are bad, because they perpetuate the harm.

I argue that paranormal services are better provided by people who understand their limitations, rather than by those who believe they can do something they can't. In fact, if paranormal services were regulated, this would be the law. Think how much better off believers would be if the paranormal services they received always led them to trained professionals in cases where such is needed.

However, these cases are in the minority. Most of the time, people who buy paranormal products or services — be it goddess worshipping seminars, homeopathy, acupuncture, or psychic readings — are buying completely harmless services that P.T. Barnum would have been happy to sell. If money is changing hands, and responsible adults are going into it with their eyes open, they receive exactly what they want, and they are completely satisfied with the results, then I would have no problem participating in such a transaction and profiting from it. The customer is happy, the peddler is happy, nobody is hurt, everybody involved is enriched by the transaction. This is their choice. They don't have a problem with it, why should you? It's none of your business.

5. SUSTAINABLE SUSTAINABILITY

I bet you didn't know that this is a sustainable book, delivered on sustainable paper, using sustainable printers, and read by your sustainable eyes. Now you know. But really you should have known that already, because this year's winner of the meaningless, overused buzzword award has to be the word "sustainable".

To label your product as "sustainable" is to imply that competing products are not sustainable. What this is intended to mean is often pretty vague. Presumably it means that competing products are manufactured from materials that we'll run out of, should current methods and usage continue.

The environmentalists, usually portrayed in the media as the good guys, first coined the phrase to describe products or methods that are generally better for the environment than the competition. Soon the marketing gurus got ahold of the word, and now everything from toothpaste to music to real estate is being sold as "sustainable".

It's so effective, and thus popular, because it's an alarmist term. Calling your product sustainable is not really saying anything your product; it's clanging the warning bell about the alternative being unsustainable: Can't be sustained! The world is ending! It's like calling your product "hate free" or "cruelty free". In no way is it descriptive of your product, it's simply an underhanded way to insult your competition. As any marketing expert will tell you, people respond much better to a negative than to a positive.

One gross overusage of the term is "sustainable agriculture", used almost exclusively by those selling organic crops. Organic agriculture is certainly sustainable, so long as a third of the world's population is willing to die off so the rest of us can eat. As with many people who use the word sustainable, proponents of organic foods aren't really saying anything particular about their product, they're trying to frighten you into thinking that modern advanced farming methods will somehow destroy or deplete the environment, and are thus "unsustainable". Ironically, the reverse is closer to the truth. Among other benefits, modern hybridized crops are designed for specific soil types, and to leave those soils less depleted so that they can be replanted for more seasons before being rotated. So-called sustainable agriculture is, in fact, far less sustainable than the planting of crops that have been optimized to thrive in the available conditions.

The word "organic" is itself the same kind of deceptive marketing: intended to trick you into thinking the alternative is somehow not organic. Strictly speaking, all plants and animals are organic, according to the word's true definition. When you hear any product defined only by a vague buzzword, be skeptical.

You also hear a lot about sustainable fuels for cars. This usually refers to biodiesel and ethanol, since they come from renewable resources instead of a limited resource, natural petroleum. In this sense, the production of biodiesel and ethanol is certainly more sustainable than gasoline, since we'll always be able to grow them. However, they have a show-stopping drawback. Burning biodiesel or ethanol in our cars exhausts the most significant greenhouse gas, carbon dioxide, into the air — just like gasoline does. So even if we switched all of our cars over to biodiesel and ethanol tomorrow, down the road we'll be no better off. The production of biodiesel and ethanol might be sustainable, but their usage is absolutely not. This is a great example of why you need to bring a skeptical attitude when you hear the word "sustainable". Are the environmentalists promoting biodiesel really looking out for what's healthiest for the earth, or do they have some other motivation, possibly political, possibly economic, possibly philosophic?

The word sustainable has become so pervasive that its usage is often just plain silly. Colgate recently purchased a company that makes sustainable toothpaste. It contains bone powder. Does an intelligent person really think that it's unsustainable to make toothpaste any other way?

Sustainable tourism is being marketed everywhere. It usually describes destinations where the attractions are generally undeveloped, like the Amazon. It is really unsustainable to vacation in developed destinations like Paris or Tokyo?

Sustainable economics are particularly bizarre. Google the term, and you'll find that it's used largely to refer to wealth redistribution. Has communism really proven to be more sustainable than capitalism?

A prominent automotive magazine recently tested four "sustainable sport sedans". Are four cars that get marginally better gas mileage than other similar cars — none of which are particularly great — honestly the only type of vehicles whose production can be sustained?

Sustainable music is also all over the Internet. In one case, it means the guy makes his own instruments. Is "sustainable" really the word that best describes that? Playing an instrument someone else made is not sustainable? In other cases, it refers to songs about anticorporatism. Is it truly impossible to sustain the playing of music about other themes?

I found a web site offering sustainable real estate. Two of the houses were built of corn cobs and hay bales (I wish I was making this up). I'll ask the Big Bad Wolf how sustainable that type of engineering is.

There's no doubt that doing things in a truly sustainable way is good. Accomplishing a worthy goal in a way that's infinitely repeatable is best, and that's what sustainable really means. True sustainability might violate the laws of thermodynamics, but we'll cross that bridge when we come to it. It's still a good goal, and as such, sustainability deserves not to be diluted into a meaningless buzzword. Thus, true environmentalists should be the first ones to object to the misleading pop-culture usages of the word that we see every 2 minutes. When you hear it, be skeptical. Figure out what they're really trying to say, and what their motivation is. And for God's sake, don't buy any bone-powder toothpaste just because it says "sustainable" on the package.

6. WHEATGRASS JUICE

The other day, I was getting a pineapple smoothie for lunch, when I happened to notice a poster extolling the virtues of wheatgrass juice. I didn't know too much about it, except that I've heard a lot of people talk about it as if it's the second coming. So out of curiosity, I began reading.

My friends, the English language does not contain adequate hyperbole to do justice to the tons of manure printed on this poster. If the average person can take even half of this poster seriously, then the ignorance and gullibility of the general public is much worse than even I would have ever guessed.

The poster was a list of claims, almost all of which centered around chlorophyll, of which it said wheatgrass juice is a rich source. Chlorophyll, as you know, is used by plants to synthesize proteins and sugars, using radiation from the sun to power a chemical reaction, converting carbon dioxide from the air and water from the ground into proteins and sugars, exhausting oxygen as a waste byproduct. Humans and other animals, not surprisingly, don't work this way. We get our proteins and sugars by eating food; our bodies have no special use for chlorophyll.

Now, I'm not saying that there's anything unhealthy or bad about wheatgrass juice. It's probably at least as healthy as just about any other plant that you could put in your juicer and blend into green syrup. I probably wouldn't rate wheatgrass as high on the nutrition scale as a proper vegetable, but I doubt very much that there's any harm in it. However, wheatgrass juice proponents don't merely claim that it's healthy. They've assembled the most outrageous list of vague medical conditions that it cures, and all sorts of types of wellness that it supposedly promotes. Since these claims are all entirely unsubstantiated, and sound far fetched to say the least, this is certainly a product you should approach skeptically. Let's take a look at some of these claims.

Wheat grass is high in oxygen like all green plants that contain chlorophyll, and the brain functions at an optimal level in a high-oxygen environment.

While it's true that if you cut off the oxygen supply to your brain, its function will be somewhat less than optimal, it's not true at all that chlorophyll is a good source of oxygen. I suggest you continue to rely on your lungs for that, which are probably better, since you don't have leaves. All types of chlorophyll have only trace amounts of oxygen. Chlorophyll is a carbohydrate, thus its makeup is overwhelmingly carbon and hydrogen. The molecule has as many as 127 hydrogen and carbon atoms, but only 5 or 6 atoms of oxygen, 4 of nitrogen, and one lonely magnesium atom. Incidentally, this also refutes another claim: that the high magnesium content of chlorophyll builds enzymes that restore your sex hormones. Interesting, given that enzymes are proteins made of amino acids, which contain no magnesium at all. I have no idea whether a single atom of magnesium restores sex hormones, whatever that means, but if so that's one hell of an atom. If you want magnesium, take a vitamin pill. If you want oxygen, take a breath. If you want sex hormones, get a girlfriend.

Wheatgrass juice has been proven to cleanse the lymph system, building the blood, restoring balance in the body, removing toxic metals from the cells, nourishing the liver and kidneys and restoring vitality.

The grammatical errors are from the poster, not from me. Let's take these one at a time. First, the claim that it's been "proven" to do any of these things. Notice that these claims are very vaguely worded: "restores balance", and "builds the blood". This is deliberate. If they tried to be specific, they would get into trouble with the FDA. If you make a claim that a product is intended to diagnose, mitigate, treat, cure, or prevent a specific disease, and your product has not been tested and is therefore unregulated, you're in violation of the law (21 U.S.C. 343(r)(6)). The wheatgrass people of course can't actually submit their product for testing against any particular diseases, because of course it would fail. So they are relegated to making only vague, untestable claims like it "builds the blood" and "restores balance".

As for whether the ingestion of wheatgrass will lower the levels of toxic metals in your cells, I wasn't able to find any research that supports this. However I did find research where living wheatgrass was found to be one of the grasses most susceptible to the absorption of zinc and cadmium from the soil through its root system, so it's more likely to be contaminated with these metals. If lowering your toxic metal levels is important to you, wheatgrass is the last thing you should put on your shopping list. And, of course, this is all founded on the assumption that you have toxic metal problems that need to be addressed. Before you seek out quack remedies for this unusual illness, first find out from a medical professional that this is indeed a problem you have, and don't go only on the assurance of a health food store owner who wants to sell you something. Probably the best thing you can do is stop chewing the lead paint off your windowsills.

It contains most of the vitamins and minerals needed for human maintenance, including the elusive B12.

Sounds compelling! But it sounded less compelling when I turned away from the poster, and looked at the store's own nutritional facts chart. The only vitamins present in a 2-ounce shot of wheatgrass juice are 15% of your daily allowance of Vitamin C, and 20% of iron. The rest of the vitamins and minerals, "elusive B12" included? Zeros, all the way down the board. The bottom line is that a shot of wheatgrass juice offers far less nutrition than a single Flintstones vitamin pill.

I would welcome a scientific test of wheatgrass juice. I challenge wheatgrass proponents to pick any supposed benefit of wheatgrass juice, and substantiate it in a real test. And by a test I don't mean a report from a hippie whose energy fields have been rejuvenated. I mean one of their claims that some sick people might actually believe and are avoiding important medical treatment as a result, such as the claim that wheatgrass juice will reduce high blood pressure. That's easy enough to test in a real, peer reviewed, double blind clinical trial. Take notice that the wheatgrass proponents have not done such a test, and there's probably a very good reason they've avoided it. Approach far fetched claims with skepticism, especially those that have not been, or cannot be, substantiated.

In the meantime, I'll continue to enjoy my pineapple and banana smoothies, no wheatgrass juice, bee pollen, or extract of ginseng needed.

7. Pond Magnet Foolishness

With a net in one hand and a pH testing kit in the other, let's wade into the murky waters of pond chemistry to test the latest fad in koi pond maintenance: magnets.

I was visiting my cousin up in Portland Oregon, and he showed me his cool koi pond. Being a koi pond guy myself, we compared notes on filter equipment, water testing, plant types, and all the usual stuff. His main filter pipe had a group of powerful magnets arrayed around it, which was something I hadn't seen before. I'm no super expert on ponds, so I guessed that maybe it was doing something like grabbing out metal filings from the pump. I asked him what the magnets were for, and he wasn't sure, but it was something to do with water chemistry. Right away, my radar went up. Unless there were significant amounts of iron, nickel, or cobalt in his water that required being magnetically held against the side of the pipe, there's really no physical way for the magnets to have any effect on anything in the water.

My cousin's friend at the pond store had recommended that he install the magnets, and he'd followed her expert recommendation. After all, he had no reason to doubt her suggestions. At my urging, he called her up to ask what the heck the magnets were supposed to do. She hemmed and hawed, said something about water clarity or chemistry or algae, and finally confessed that she had no idea, and that it was just a standard thing that a lot of pond owners do. The magnets were pretty expensive, so it wasn't surprising that a pond store would push them.

So I turned to the Internet, as I often do in times of need. It didn't take much searching to find the standard claim about magnets and ponds, and it has to do with algae growth. The claim is that magnets, mounted inline along any of the pipes, improve water clarity by altering iron alignment in free-floating algae, thus inhibiting photosynthesis. I also found one or two references to reducing lime scale build-up inside the pipes, but since this claim didn't even pretend to suggest a mechanism that might produce this effect, I discounted it. Lime scale is calcium carbonate. It contains none of the three magnetic elements and is thus completely unaffected by magnets. That claim pretty much busts itself, no help from me needed.

So what about this reduction of algae? The claim is that the algae will be reduced because its photosynthesis will be inhibited, due to the realignment of its iron. This is a fairly common type of claim. It makes no sense, but because it uses common scientific-sounding words, many people will simply accept it at face value without questioning it. My cousin's friend at the pond store did, and when she repeated it to my cousin, he did too. I even accepted it when he told me, albeit tentatively, pending some kind of reasonable explanation.

Here are the two problems with this claim. Number one, photosynthesis is a chemical reaction among carbohydrates. Iron is not involved. The presence of iron would neither hinder nor help photosynthesis. The magnetic orientation of any iron molecules nearby is not relevant. Realigning iron magnetically has no bearing on photosynthesis, and will not harm a plant in any way. Number two, iron, which is found in human blood hemoglobin, is *not* present in chlorophyll or in the other proteins involved in plant photosynthesis. Although I've never spent the time to wave a magnet past a plant several times a day, I'd be awfully surprised if that plant's photosynthesis stopped and it died as a result.

Another problem with this claim is the concept that briefly passing a non-magnetic object through a magnetic field will leave it altered after the magnet is removed. This is like turning the light in a room on then off again, and expecting the furniture to be somehow residually contaminated with light. Electromagnetic radiation doesn't work that way.

I should mention that when I set out to research this claim, I didn't merely gather enough information to shoot the claim down and then quit. I did make a good effort to find research supporting the effects of magnets on algae. But, since there are no plausible claims, there has never really been anything for anyone to test. However, I did find something close. In 2005, the Woods Hole Oceanographic Institute announced the results of research they'd done on certain bacteria that are known to carry magnetic crystals. These are called magnetotactic bacteria. In 1970, magnetotactic bacteria were also discovered in the Southern Hemisphere whose magnetic crystals were flipped around. The purpose of these tiny internal compasses has never been known, but since the 1970 discovery, the working hypothesis has been that they use the compasses to help navigate either up or down to find water with the best oxygen concentration. This would be consistent with the need for the polarity to be reversed in the Southern Hemisphere.

Alas for the pond magnet manufacturers, Woods Hole's research found that north-polarity and south-polarity bacteria are both found intermixed in both hemispheres, and also that there are numerous individuals who lack the crystals completely. All three types of bacteria navigate equally well to the water depths with the most desirable oxygen levels. The conclusion of the research is that the purpose of the magnetic crystals remains unknown, but it's clear that its reversal or even its total lack makes no difference to the health or life cycle of the bacteria. And, once the bar magnet was removed from the microscope slide, the magnetotactic bacteria realigned themselves normally with the earth's magnetic field, according to the polarity of each. There were no residual effects of having been briefly placed near a magnet.

I recommend that you do your own research, or at least ask for a reasonable explanation, when a salesman offers you a product that claims to accomplish something far fetched or contrary to your understanding of the laws of nature. This rule should be considered only a guideline and can be relaxed, when the salesperson is (like my cousin's friend) a really hot chick.

8. Nocturnal Assaults: Aliens in the Dark

I was five years old when my single mother was attacked by a ghost in her bed in the middle of the night. She awoke suddenly under the pressure of two unseen hands pushing her down flat against the bed and holding her there. For several minutes she struggled, unable to speak or move. Finally she broke free and scrambled out of the room, and spent the rest of the night on the floor of the room that my brother and I shared. She never went back into her own bedroom alone again. And so I grew up with this history, hearing ghost stories from time to time that other people told, but knowing that we had actually had a real ghost in our home when I was young.

I was an inveterate reader of books about monsters and ghosts — everything from Bigfoot to Dracula, from banshees to fairies, from zombies to werewolves — and one subject that particularly piqued my interest was that of nocturnal assaults. Nocturnal assaults are attacks just like that suffered by my mother, though often more graphic: the attacker can sometimes be a visible apparition. I was highly intrigued to learn that the physical descriptions of the attackers have been eerily similar over the ages, varying by country and sometimes by century. In Anglo cultures the most common attacker is called the Old Hag, a terrifying old woman dressed in black rags who holds her victims down in their beds or even sits on their chests with her full weight. References to the Old Hag and her nocturnal attacks go back as far as the Middle Ages. She's been part of our history for so long that if you haven't slept well, you're said to look "haggard". In India she is the Mohini, a beautiful but deadly enchantress. As often as the Anglo attacker is described to look like an old hag, attackers in India are just as frequently described as a beautiful young woman with terrible powers. In Slavic cultures, the most frequent description is of an elf-like gypsy man with wild glowing eyes who sits on your chest, riding you like a horse. The more I researched it, the more cultural groups I found to have their own unique nocturnal assault perpetrators.

As a budding young scholar of the supernatural, I was fascinated by these cultural commonalities. Similar attacks, throughout history, made by specific attackers who stayed within their own cultural communities. And then I had a breakthrough. Beginning in the late 1960's, a new attacker began muscling in on the Old Hag's territory, and quickly took over responsibility for most of the attacks reported in the United States. Do you know who I'm referring to yet? In 1965, Betty and Barney Hill went public with an episode they said happened to them in 1961, when they were abducted from their car by aliens, and suffered terrible medical experiments aboard a spacecraft. Curiously, the attack they described bore no resemblance to a classical nocturnal assault; however the creature they described — an alien of the type we commonly call a "gray" — became America's new supernatural superstar. Nocturnal assaults continued to happen at the same frequency that they always had, but now the reported attacker was, more often than not, a gray alien. The gray alien burst upon the scene of America's consciousness just as the Old Hag was beginning to seem a little outdated and, well, haggard. Just as children in India grew up with stories of the Mohini as the evil specter who might paralyze you in the middle of the night, we're now in a generation of Americans who have heard that gray aliens are those little beings who are going to come into your bedroom at night and attack you.

Is it really as simple as that? Is the attacker that your scared brain visualizes based solely on what your cultural experience tells you to expect?

It was about 25 years after my mom's attack that I first heard of sleep paralysis, which, as you probably know, is the clinical name of these nocturnal assaults. Sleep paralysis can be characterized by an inability to speak or move, a feeling of intense crushing weight on the chest, and/or hallucinations which can be visual, auditory, tactile, or even strange smells. It happens only during REM sleep, often just as it's beginning or ending. Sleep paralysis is five times more likely to happen to people sleeping on their backs, facing up. Drugs such as Prozac have been found effective in controlling sleep paralysis attacks. Although most sleep paralysis episodes do not include the visual apparition, more than enough do include it to account for all reported nocturnal assaults. Sleep paralysis is well understood, well documented, and is an accepted psychological phenomenon among almost all medical professionals.

So why, then, did it take me a further several years before I made the connection between my mom's attack and sleep paralysis? I had spent so many years fully believing that my mom had been attacked by a ghost that it never even occurred to me to seek more reasonable explanations elsewhere, even when the obvious answer was staring me in the face, literally, as I was reading books about it. Perhaps this is the same reason that even in an age where most people have at least heard of sleep paralysis, believers in alien abductions and nocturnal ghost attacks firmly stick to paranormal explanations for their own sleep paralysis experiences.

Many believers, when confronted with this explanation for their experience, will point out differences between their experience and the known symptoms of sleep paralysis. Of course, visual, audible, and tactile hallucinations are part of the known symptoms of sleep paralysis, so it's kind of hard for them to come up with details that can't be attributed to known sleep paralysis effects. And that's an uncomfortable position to be in as a skeptic: no matter what the believer reports, we can explain it with "It's a hallucination." That's just like creationists being able to explain anything with "God did it," no further evidence needed. The difference is that we can actually test nocturnal assault sufferers, and whenever we do, we end up with video of them lying in their bed looking paralyzed, with a conspicuous absence of gray aliens in the room.

So it took over thirty years, but I finally did explain my mom's nocturnal assault, at least to my own satisfaction. You might wonder what her own assessment is, in light of this explanation. She went to medical school, spent her whole career in biotech, has a very scientific mind, and is convinced to this day that she was attacked by a ghost. She never read the Betty and Barney Hill story.

9. SIN: WHAT'S IT GOOD FOR?

Now I'm going to put on my burgundy velvet robe, fill my martini glass, and observe that bastard stepchild of the value system: Sin.

Sin is an interesting thing. A sin is something you're not supposed to do, according to a given set of religious restrictions. Sins are not necessarily illegal. Sins are not necessarily wrong. Sins don't necessarily harm anyone. In fact, many sins can be completely, entirely harmless, like the thinking of impure thoughts. So what's the problem? Why are sins bad?

I guess that all depends on whose definition of "bad" you use. For example, if you're a Muslim, it's sinful to get urine on yourself. The rest of us follow this commandment pretty strictly too, but we certainly wouldn't consider the odd dribble to be sinful. Buddhists consider skeptical doubt to be a sin (though they call it a hindrance), but doubt certainly isn't a problem for Christians or Muslims. Most Christians consider polygamy to be sinful, but it's the rule for most of Africa and the East. So there's no one clear yardstick for determining what's sinful or not. It depends completely upon the religious context. Outside of a religious context, the word sin is, for all practical purposes, meaningless.

Christians in particular consider everyone to be sinful, regardless of their performance. They call this "original sin", and it's essentially a negative blot on your report card immediately upon birth. Since Adam and Eve had the gall to eat some fruit that was offered to them, you and I and everyone else are considered guilty by association and are thus fundamentally bad people, according to strict Christian doctrine.

Christians also have to deal with "mortal sins." A mortal sin is one that, if left unrepented, sends you to hell when you die. Christians don't maintain a list of what types of sins guarantee you a date with the devil, instead they lay out some general rules. The big sins, like murder and adultery, put you on the fast track. Mortal sins have to be done deliberately. If you simply forget to go to church, accidentally put on a condom, or unintentionally catch a glance at a hot chick out of the corner of your eye, such sins are called venial sins and you can get away with them. But if you do them deliberately — blow off church on purpose in order to saw some extra logs on Sunday morning, wear the condom on purpose, or deliberately stare at the hot chick with impure thoughts — they are mortal sins. If you do things like this regularly, strict Christians consider that you are hellbound for sure. There are probably a lot of human males who needn't bother wearing their jackets for their burial.

Worst of all is the "eternal sin" - to deny God, which cannot be forgiven. Those considering an eternal sin might as well lose a fiddling contest to Satan right now. The punishment for an eternal sin is the same as for a mortal sin; the difference is that there's no opportunity to be forgiven and get out of it. It's sort of like being on death row in a state where the governor doesn't have a telephone.

When you eliminate activities that injure others or are otherwise wrong, there are still items on the sin list: basically a long list of victimless crimes. This is where the fun begins for those of us not hampered by religious restrictions.

Take social relationships, including plural marriages, same sex marriages, and anyone living together or having sex outside of wedlock. It doesn't hurt anyone, everyone involved has a great time, and it's mutually fulfilling for all participants. But those activities are all pretty high on the sin list. Take it out of a religious context, and suddenly there's nothing wrong with it. Polyamory is also a victimless crime that for some reason is considered sinful: wife swapping, swinging, hedonism, group sex parties, and open marriages are things that all the participants enjoy behind closed doors. Where's the harm?

Straight sex between married partners is all right, so long as it never extends to include masturbation, fetishism, lust, or impure thoughts. "Have to stop a minute, Mabel, I started to feel a little lusty."

The list of sins is not static: it's even been updated to include cybersex. Using a computer in some way to enhance sexual stimulation is sinful. This includes a video chat session with your spouse when one of you is traveling. That makes a lot of sense.

Drunkenness and tobacco are big on the sin list. This one's just plain counterproductive. Who among us doesn't appreciate an evening at the club in an overstuffed leather chair, with a martini and a fine cigar, talking politics and blasphemy. Throw in some profanity (which, fortunately, I don't see on the list of sins), and you've got the perfect evening. Drunkenness and tobacco are fundamental to healthy male adulthood. Frankly, I don't even know how I'd be able to conduct a proper board meeting without these accoutrements.

Idolatry is another sin that would be hard for me to live without. Idolatry doesn't necessarily relate to graven images or statues of other gods; idolatry is the practice of loving anything or anyone more than you love God. For me, the brand names Porsche and Jeep are hard to get past. I do attend church every Sunday morning: My temple of worship is a rectangle at the beach measuring 8 meters by 16 meters and involves the hitting of a synthetic leather ball at other worshipers. And since I cannot honestly say that there are any supernatural invisible flying magicians whom I love more than my own family, idolatry is definitely a sin that I need to commit every minute of every day, as much as I need to draw breath.

Hate and anger are sins. I don't really hate anyone and I don't get angry very often. About the only thing that gets me angry is when I hear the worst of the bad news from the world: children being abused or murdered, and genocides. Apparently, the world's major religions think that I should go to hell because those things make me angry. I'd have to say this is one case where the world's major religions can kiss my ass. I respect how the Amish can overlook these crimes and offer loving forgiveness to even these criminals, but I'll save my applause for the inmates who beat Jeffrey Dahmer's head to death against a prison toilet.

Lying. This one's tough. I don't know how anyone can claim that they don't practice this sin every day, no matter how religious they are. Have you ever told anyone that you can't go somewhere, or can't do something, when the truth is you simply didn't want to? You're a liar. You ever stop talking about someone when they entered the room, to deceive them into thinking you weren't talking about them? You're a liar. Ever give someone one of those quick fake smiles when you pass them in the hall — as if seeing them makes you happy? You're a liar. Lies don't have to be spoken and they are usually not malicious, but they're still lies. We all do it, all day, every day. Lying is a fundamental of politeness and a pillar of good behavior.

The truth is the concept of sin has no place in the lives of intelligent adults in modern society. Politeness, honesty, industry, and simply being yourself will take you a lot further. I say to the religious people: Keep your arbitrary restrictions, and your hateful belief that I should go to hell, to yourselves.

Perhaps my favorite aspect of being free of the restraints of sin is that I get to make the following invitation: If any hot women out there want to join my wife and I for some raging topless hot tubbing, well cooled with margaritas and free of any Catholic guilt, give me a call.

10. AN EVOLUTION PRIMER FOR CREATIONISTS

Some creationists may be concerned that some of their standard arguments against evolution sound dismissive or patronizing. This is probably true: in any debate, it's common to frame your opponent's arguments in a weak light. Sometimes this is done deliberately to make evolution sound ridiculous, and sometimes it's done accidentally through ignorance of what evolution is and how it works. Since misinformation and ignorance are poor platforms on which to build any conversation, I present the following Evolution 101 Primer for the benefit of creationists who want a correct basic understanding of their foe. I think the best way to do this is to dispel the three most common evolution myths.

Myth #1: Men evolved from apes.
This is the oldest and wrongest misconception about evolution. Nobody has ever suggested that one existing species changes into another existing species. Some criticisms of evolution show illustrations that fraudulently purport to show what evolutionists claim: that a salmon changed into a turtle, which changed into an alligator, which changed into a hippo, which changed into a lion, and then into a monkey, and then into a human being. Of course such a theory would seem ludicrous. But it's pure fantasy and has nothing in common with real evolution.

The diversification of species is like a forest of trees, sprouting from the proverbial primordial soup. Many trees die out. Some don't grow very tall. Some have grown a lot over the eons and are still growing today. Trees branch out, and branches branch out themselves, but branches never come back together or combine from two different trees. The path of a species' evolution is shaped like the branch of a tree, not a donut, not a figure 8, not a ladder. To embrace evolution, you need not —

must not — think that a salmon turns into a zebra, or that an ape turns into a man. It's simply not genetically possible.

We've all seen the other famous illustration, where a monkey morphs into an ape, that morphs into a caveman, that morphs into *Homo sapiens*. If you climb back down the tree branch, you will indeed find earlier versions of man where he was smaller, hairier, and dumber, but it won't be a modern ape. To find a modern ape, you'd need to go even further down the tree, millions and millions of years, find an entirely different branch, and then follow that branch through different genetic variants, past numerous other dead-end branches, past other branches leading to other modern species, and then you'll find the modern ape. Never the twain shall meet.

Myth #2: Evolution is like a tornado in a junkyard forming a perfect 747.

This is a popular manifestation of the argument that evolution depends on randomness, and so it would be impossible for complex structures to evolve. Well, this is half right, but completely wrong in its totality. Random mutations are one driver of evolution, but this argument completely omits evolution's key component: natural selection.

Obviously, in reality, if a tornado went through a junkyard, you'd end up with worse junk, not a perfect 747. No evolutionary biologist, or any sane person, has ever claimed that you would. It's ridiculous. The tornado is meant to represent the random element of evolution, but genes don't mutate catastrophically all at once, like a tornado. Here is a more accurate way to use this same analogy.

Imagine millions of junkyards, representing any given population. Now imagine a group of welders, who walk carefully through each junkyard, twisting this, bending that, attaching two pieces of junk here, cutting something apart there. They do it randomly and make only a limited number of small changes. Sometimes they don't change anything. This is a far more accurate representation of how genes mutate within an organism. It's not a single cataclysmic tornado.

Now comes the natural selection. Let's test every piece of junk in every junkyard. Does anything work better? Does anything work worse? With millions of changes in millions of junkyards, it's inevitable that there will be some improvements somewhere. Part of natural selection is the eventual removal from the population of any organisms that are less well adapted, so to simulate this, we're going to eliminate all the junkyards where the junk was worse after the welders made their mutations. This leaves only junkyards that are stable, or that are improved. To simulate the next generation of the species, we replicate all of our current improved gene pool of junkyards, and again send in the welders. They make a few random changes in each, or no changes at all.

Each time this entire process happens, the population of junkyards improves. But this doesn't happen just a few times. It happens millions or billions of times. The changes made by the welders are countless. The vast majority of changes are either useless or make things even junkier. Since natural selection automatically filters out the poorly adapted junkyards and rewards those rare improved junkyards with additional procreation, our population of junkyards gets better and better. Things start to take shape in the junkyards. Useful things. Stronger things. Things with abilities that nobody could have predicted. Any given piece of junk that improves is replicated in many junkyards, and reappears in millions of slightly altered forms each time. Pick the best version from each generation, and you can literally watch the same piece of junk evolve into a better, stronger, more useful, and better adapted machine with more capabilities. This is evolution.

Myth #3: Evolution is just a theory.

First of all, if you believe that most biologists consider evolution to be "just a theory", you're behind the times. Almost all biologists call it a fact, and not because they feel any particular need to respond to creationists.

Second, when creationists try to put evolution down by dismissing it as "just a theory", they're actually acknowledging its scientific validity. To understand why, it's necessary to understand exactly what a theory is. When creationists use the term to disparage evolution, they really should be using the word hypothesis. A hypothesis is a provisional idea, a suggested explanation that requires validation. Evolution is well beyond that stage, though; even the staunchest anti-evolution creationists assign evolution the much higher status of theory.

In order to qualify as a theory, evolution had to meet the following criteria:

❖ A theory must originate from, and be well supported by, experimental evidence. It must be supported by many strands of evidence, and not just a single foundation.

❖ A theory must be specific enough to be falsifiable by testing. If it cannot be tested or refuted, it can't qualify as a theory.

❖ A theory must make specific, testable predictions about things not yet observed.

❖ A theory must allow for changes based on the discovery of new evidence. It must be dynamic, tentative, and correctable.

Notice that last one: tentative, correctable, and allowing for future changes. Creationists often point out that the theory of evolution is incomplete, like any theory, as if this disproves it. To be a theory, evolution must be incomplete by definition, and (no pun intended), constantly evolving.

The strict scientific definition of a fact is both simpler and hazier. A fact is a verifiable observation, and evolution is verified so many times throughout the entire science of biology that most biologists call it a fact. However many scientists contend that every fact has some element of theory to it, so in this sense, it doesn't really make any difference whether evolution is called a fact or a theory. Since biologists are always learning more and adding to our knowledge of evolution, it's probably best to leave it as a theory.

I hope some creationists find value in these explanations.

11. Killing Bigfoot with Bad Science

In this chapter, we're going down a dark forest path on the trail of *Gigantopithecus americanus:* the North American Sasquatch.

I see many cases on both sides of the Bigfoot debate where bad arguments, bad science, and just plain weirdness is being put forth, doing great disservice to their own side of the argument. There are intelligent and productive ways to explore a subject and present a case, but I don't see it being done very often on either side of the Bigfoot debate. I'm going to present what I consider the top three ways that each side of the Bigfoot claim is shooting themselves in the foot, beginning with the skeptics.

1. Saying that the guy who confessed to making tracks disproves the entire thing.

In 2002, a Washington logger named Ray Wallace died, and his family produced the carved wooden feet that he used to make Bigfoot footprints all over the Pacific Northwest, beginning in 1958. The newspapers and TV tabloids lapped it up, reporting that the entire Bigfoot phenomenon was now proven to be a hoax perpetrated by Wallace. Well, I feel the time has come for me to come clean about something that I've wanted to get off my chest for decades. When I was a kid, I once made some fake Bigfoot footprints too. The cat's out of the bag. Bigfoot is now doubly proven to be a hoax.

Obviously, anyone who has any kind of basic understanding of research methodology can't accept Ray Wallace's story as proof that Bigfoot is a hoax. Sure, he made fake prints. So have a thousand other guys. They were doing it before Ray Wallace was born, and they're still doing it today. Anyone can be making those tracks. Anyone...

2. Saying the Patterson-Gimlin film is "the worst fake ever."

I'm not a Bigfoot believer but I will give credit where credit is due. The Patterson-Gimlin film looked like a real animal to me. The Discovery Channel's "duplication" of it looked ridiculous. It looked nothing like a real animal, and certainly didn't remotely resemble the subject shown in the Patterson-Gimlin film. Chewbacca looked more real than the Discovery Channel's Bigfoot suit. Hollywood's state of the art in gorilla suits in 1967 were *Planet of the Apes* and "The Galileo Seven" episode of *Star Trek*. Two loggers with no previous gorilla suit experience made a suit that was better than today's state of the art, and certainly light years ahead of the 1967 state of the art. I'm not saying the film's real, I'm saying give credit where credit is due, and admit that if it is a fake, it's astounding. If you disagree then go through a stabilized version frame-by-frame as I have.

The half dozen or so Hollywood special effects artists who have since "come forward" to claim that they were responsible for the Patterson-Gimlin Bigfoot suit, and the dozens of guys who have "come forward" to claim that they were the guy wearing the suit, are no more evidence against the film than Ray Wallace's wooden feet are evidence that no real Bigfoot footprints exist.

Critics of the film also say that the creature's behavior is unrealistic. I have no knowledge of what a real Bigfoot's behavior might be, but I have encountered bears half a dozen times, and they acted exactly like the Patterson-Gimlin creature: just walked away, unconcerned, with maybe only a look or two back.

3. Criticizing good scientists like Jeff Meldrum.

I've read old and new criticism of Dr. Jeff Meldrum of Idaho State University, and I'm only mentioning his name in particular as one example. There are several prominent tenured professors at legitimate accredited universities who have done Bigfoot research. They are probably far, far outnumbered by professors who have done psychic or other paranormal research, but let's stick to the subject.

Dr. Meldrum is not the obsessed Bigfoot guy who lives and breathes it 24 hours a day, and exhorts his students to become believers. Rather, he has a long list of publications and edited volumes, none of which pertain to Bigfoot; he teaches six courses, none of which pertain to Bigfoot; he's an Associate Professor of Anatomy & Anthropology; he's an Adjunct Associate Professor in the Department of Anthropology and the Department of Occupational and Physical Therapy; and he's the Affiliate Curator at the Idaho Museum of Natural History. He's not the Professor of Bigfoot. He carries as great a load of academic work in non-Bigfoot related studies as any professor. He's a real scientist doing real work. On top of all of this, he studies casts of Bigfoot footprints.

Dr. Meldrum is responsible for drumming up his own grant money from private donors to fund any Bigfoot research that he chooses to do. In some cases, he has received small amounts of matching funds from the university. If you feel this was a bad expenditure, then criticize the university regents who decided to write the check, don't criticize the person they gave the funds to. The work of responsible scientists like Dr. Meldrum is exactly what true skeptics should be asking the Bigfoot community for, not criticizing him for it.

Here is the way for a responsible skeptic to handle the Bigfoot claim. It's to say "You're making an extraordinary claim. Show me extraordinary evidence, and I'll believe it. Until then, I'm not convinced." Occasionally candidate evidence has come forward, like hair and stool samples, or the skull cap from Tibet. This evidence has been properly tested, and so far no new great ape species has been proven (and if I'm wrong about that, I invite your comments on the web site). A responsible skeptic's obligations do not extend to poking fun at the people who are looking for evidence, considering the lack of evidence to be proof of no evidence, or making personal comments about people. That's not good science. In some cases, Dr. Meldrum, and other scientists like him, are being better skeptics than the skeptics.

And now, I'd like to say a few words to those who mean to support Bigfoot but do themselves more harm than good with bad arguments. The wrong ways to support Bigfoot:

1. Stating that Bigfoot is an extraterrestrial, or comes to us from another dimension.
If Bigfoot claims are going to make any headway into mainstream science, it will be through zoological channels, not supernatural channels. Such claims are the most extreme form of counterproductivity, setting Bigfoot claims backwards all the way into the Dark Ages.

2. Being delusional: Seeing detailed Bigfoots in a blurry photograph that shows no such thing.
Half the Bigfoot web sites out there show numerous photographs of bushes and wooded areas, with certain areas circled. There's nothing within the circled area except other bushes; maybe a shadow, or a dark branch. But wait! Here's a detailed sketch of what's hiding inside that shadow. I'm not a psychologist so I won't presume to affix a label to this phenomenon; but seeing things in pictures that aren't there, and then obsessing over it, does not strike me as healthy. It's certainly more effective at raising concern for the claimant, than it is at convincing anyone that Bigfoot exists. If all you have is bad evidence, you're better off not presenting it.

3. Doing bad science: Seeking to support a preconceived conclusion.
Science doesn't work by starting with the goal of proving something and then assembling whatever evidence you can find that supports it. That's doing propaganda, not science. Start with a testable hypothesis, and then form a theory based on the evidence revealed by the data. Of course, following this method is going to make it pretty hard to come up with a theory that's supportive of Bigfoot, but that's what it's going to take if Bigfoot supporters hope to prove their point.

I know you're going to read all of this and conclude that I'm the pro-Bigfoot guy. I'll admit to being a Bigfoot hopeful (a hope based more on emotion than on any actual likelihood), but certainly not a believer. My point is simply that both sides of every debate contain a lot chaff along with the wheat. Both sides of every skeptical issue believe that they're right, but even those on the side that is right (and by that, I mean whichever side *you're* on) can probably stand to clean up their act a little, no matter what the issue is.

12. KILLING FAITH: DECONSTRUCTIONIST CHRISTIANS

Now we're going to take a leap of faith into a soft cushion, to see what happens when proven knowledge makes faith irrelevant.

There is a profound contradiction rising in the world of religion. Proponents of various religious dogma such as Creationism, Noah's Flood, and Revelations have taken a disturbing turn. They are crippling their own religion by attempting to do scientific research in an effort to prove their religious claims, thus directly attacking their religion's central pillar: faith.

Abraham is regarded as the father of faith among most of the world's people, including Muslims, Christians, and Jews. He earned this title through demonstrating the mightiest act of all: being willing to sacrifice his own son Isaac, indeed with the dagger poised above his head ready to fall. Isaac was saved when God sent an angel at the last second to put a stop to it, who told Abraham that he'd proven his faith. It was an act that very few among us could have duplicated; I certainly wouldn't have done it. For this reason, Abraham is rightly exalted. It was truly an act of heroic faith.

Consider this question: If Abraham had known that God would intervene at the last second to spare Isaac, would his act have been as heroic?

Theological tradition tells us no, it would not have. The reason the Abraham story is important is that it's the supreme demonstration of faith. Abraham raised his dagger fully intending to kill his beloved Isaac, all for his faith in God. He felt every ounce of the unimaginable anguish. Could you have brought the dagger down and plunged it into your own child? Achieving this level of faith is the essential goal of all Christians, and for that matter, it is for Muslims and Jews as well. Faith is the absolute pillar of religion.

Now let's turn the clock forward a few thousand years and see where the faithful are today. Surprisingly, I see a lot of them doing the equivalent of asking questions before raising the dagger. Questions like "Can you please prove to me that the angel's going to intervene?" Can you show me the scientific evidence that proves Intelligent Design? Can you please prove to me that Moses parted the Red Sea?

The Associates for Biblical Research (abr.christiananswers.net) publishes a quarterly PDF document called "Bible & Spade". It's all about archaeological projects throughout the middle east that they say supports the Biblical record. The current issue offers evidence from Egypt on the location of the Exodus crossing of the Red Sea. They have an exhaustive mission statement page, in which they state and restate their belief that the Bible is absolutely and literally a correct and true historical document. It is "infallible, inerrant and authoritative". Their purpose also includes "Edifying the Christian Church by encouraging a deeper knowledge of, greater appreciation for, and stronger faith in the Bible through knowledge and correct interpretation of the findings from archaeology and science." In short, they are all about proving the Bible is true through archaeology. They call this "encouraging stronger faith in the Bible". Encouraging faith through proof. They want to force us to believe it.

Maybe my dictionary is out of date, but faith and proof are oil and water. Faith needs no proof, and in the presence of proof, faith becomes irrelevant. Faith means to believe without proof; indeed, it means to believe in spite of evidence to the contrary. Where is the heroic faith in believing in something that's proven right before your eyes? That's hardly a demonstration worthy of Abraham. To seek to marginalize the element of faith by showing supporting evidence, is to seek to undermine the whole basis of the religion.

We see the same thing happening in any of the numerous groups seeking to find Noah's Ark on Mt. Ararat in Turkey. On some of their web sites you'll find tremendous amounts of information about how a wooden ark could have survived 6000 odd years, how it could get so high on the mountain when there's not enough water on the planet to do it, exactly where it's located in the satellite photographs, exactly how two of every animal could fit on one ark, what its dimensions are and where and how it was built, and so forth. But nowhere did I find an explanation of why it's important that it be found. To my way of thinking, even if you're of the mindset that Noah's flood was simply a literal account of an incident and not a meaningful allegory, then allowing it to be found, thus proving the story, would be more likely to be on Satan's agenda than on God's. Why would God want to marginalize faith? I can think of every reason why Satan would want to do this, but not God.

Is proving the Bible really doing the work of God?

Abraham's faith did not need the crutch of supporting scientific evidence that God is real, nor would he have made much of an impression upon God if he'd had such. I challenge Christians who are true believers to stick with their faith, and to hold their faith to be (if I may borrow the terms) "infallible, inerrant, and authoritative". Or, if you want to use what science tells us instead, then admit that you're no longer keeping your faith in the infallibility of the Bible. You cannot do both. A true Christian must question their fellow believers who attempt to erode faith through the application of science to scripture. If faith is not enough to support religion on its own, then faith has already been killed.

13. A Primer on Scientific Testing

In this chapter we're going hide our crib notes in our hats, pull our sleeve down to cover the notes written on our arms, and dive into the world of testing.

Much of the feedback I've received on my position regarding wheatgrass juice concerned claims that wheatgrass juice has already been tested and been proven to cure many different diseases and promote many types of well being. And if I had a dollar for every email I received accusing me of being in the employ of Big Evil Corporations who are frightened of wheatgrass juice, I would be able to afford a shot of this quack elitist scam beverage every single day.

So I hereby present this primer on scientific testing, to better equip the layperson with the ability to determine the validity of claims being made, or the published results of supposed research. Valid claims and real research will follow the whole process that I'm about to outline, and they'll tell you about it too. If the poster you read in Jamba Juice doesn't detail the testing procedure used to substantiate its claims, or if the testing procedure is not similar to that outlined herein, then you have very good cause to be skeptical of any claims that it makes. If something works, its makers should be happy to prove it to you.

Testing of something in medicine, for example, is done by what we in the brotherhood call a clinical trial, more formally known as a randomized controlled trial. The same general principles apply to any kind of scientific testing. The aspect of randomization refers to the random distribution of subjects into similarly sized groups. When done thoroughly and responsibly, complicated statistical processes are used to remove any sort of bias for the assignment of subjects, and to ensure that the assignments are not known to the participants or the administrators. Make no mistake, even this apparently simple first step of testing is a thorough one, and it's this kind of comprehensive attention to detail that separates a real test from the typical anecdotal "testing" claimed by supporters of most pseudoscientific phenomena. One of the most important characteristics of a valid test is the control. Let's say your wrist hurts, and so you try acupuncture, and your wrist feels better. You're likely to consider that you've just tested acupuncture, and it worked, thus proving its efficacy. But in fact, this was not a valid test, because there was no control. Your wrist may have healed naturally. Your wrist may have been healed by a psychic in the next room. There is no way to know what effect, if any, the acupuncture had. It may have even slowed the healing, for all you really know. The most basic kind of control would have been to have at least two people with similar injuries, one received the acupuncture and the other received a control procedure, and all else would have had to be equal. With a control, you have the beginnings of a valid test.

Blinding is another fundamental of trials. Blinding means keeping the test participants in the blind. If people know what they're being given, know what results they're expected to report, or know what kind of result to look for, the results are untrustworthy. Everyone is a human being, and if you're not blinded, you may unknowingly skew the results, or you may have opportunity to wield some agenda that you might have. Blinding can be single, double, or even triple.

In a single-blind test, the participants in the experiment don't know any information that might skew the results. If they're testing a drug taken orally, the participants must not know whether they're taking the real drug or the control placebo. If they're receiving acupuncture, they must not know whether they're receiving traditional acupuncture or sham acupuncture; so in this case, participants must be carefully screened to be sure that they have no prior acupuncture experience. If they're taking wheatgrass juice, they must not be able to tell whether they're drinking real wheatgrass juice or a placebo, so it would have to be administered in some form where they couldn't tell. The purpose of blinding the participants is to prevent them from either knowingly or unknowingly manipulating the results of the test, by reporting or reacting differently.

Single blind tests are good, but double blind tests are better. In a double blind test, neither the subjects nor the people administering the tests know what group any given subject is in. They also don't know whether they're giving the real substance being tested or a placebo. A double blind test removes the chance that a test administrator might skew the results by acting differently, either knowingly or unknowingly, and thus providing information to the test subject.

Triple blind tests take it the furthest extreme. A triple blind test is just like a double blind test, but with the additional element of the statisticians also being blinded. For the people tabulating and analyzing the results of the test to be blinded, the data is presented to them in a coded form so that they're not able to know anything about any given subject or administrator. They'll see data like "Subject A was given substance B by administrator C, and had a 13% improvement." They don't know if subject A was in a control group or a test group, they don't know what substance B is, and they don't know who administrator C is. In this way they're able to present detailed results of the test that are completely unbiased, because even the statisticians themselves don't know what the data mean.

Once your testing is done, your results are ready for publication. If you want your report to be taken seriously, it needs to be subjected to — and survive — the process of peer review.

Peer review means having your research submitted to experts in the field. So who are the experts? That usually depends on who's publishing or funding the research. If it's a scientific journal, the editorial staff will usually maintain a stable of referees in the community. If it's some group considering funding your research, they'll typically have hired a panel of experts. If your research was responsibly conducted and your conclusions are well supported by the evidence, then the referees will typically give it a passing grade for publication.

Let's say a UFO researcher writes a paper that says UFOs come from another dimension, and he has some of his fellow UFOlogists — whom he considers his peers — to endorse his paper. Does that make it peer reviewed? No, because he chose the referees himself. What if the editor of an underground UFO pamphlet chooses a panel of UFOlogists who endorse the paper, does that make it peer reviewed? No, because these referees are clearly biased, and their scientific acumen would not survive any type of scrutiny from the general scientific community. Typically the publication must be one with a long standing reputation, and strict requirement, of thorough peer review. The process of peer review is not perfect, as it relies on individuals who, though they've been scrutinized by a committee themselves, are still human beings who can make mistakes, get lazy, have agendas, or just bad hair days. But peer review succeeds far more often than it fails, and if you want anyone to take your research seriously, it must be peer reviewed.

Remember: Articles that report reliable results will always detail the testing that was done and the methods used. If the claim is far fetched, and the supporting documentation of testing that the claimants are willing to share is inadequate, you have very good reason to be skeptical.

14. CELL PHONES ON AIRPLANES

Now we're going to fly up to 40,000 feet, flip open our cell phone, and call the Twilight Zone to tell them we're doing something that's supposed to be deadly dangerous.

I love *Mythbusters* and it's my whole family's favorite show, but with their episode on the cell phone ban aboard aircraft, they did a disservice to those of us who hope to get this groundless ban dropped. In case you missed it, they did a test and concluded that cell phones can potentially interfere with an aircraft's navigation system. The only instrument they tested was a radio direction finder called a VOR, which detects a radio beam coming from a ground station and points its direction. In practice, VOR is on its way out, in favor of GPS. VOR stations are each assigned a unique frequency in the VHF range between 108 MHz and 117.95 MHz, which is right above the FM radio frequency range. By contrast, the lowest frequency used by any US mobile telephones is 700 MHz; and in European mobile telephones, 450 MHz. Since the frequencies are so incredibly different, the whole debate is ended right there, for all practical purposes. *Mythbusters* used an older VOR receiver that could be tuned to receive a much broader range of frequencies, which is why they were able to detect the mobile phone signal. To be more responsible, they should have admitted that this frequency was wildly different than what any aircraft might possibly tune to. As it was, they left viewers with an inaccurate, and alarmist, impression. *Mythbusters* called the myth "plausible". No, it's really not plausible; a cell phone cannot affect an aircraft's navigation instruments in the real world, and we'll have more on that in a moment. Maybe if you threw a cell phone really hard at the GPS, you could break something.

So this raises an obvious question: why aren't cell phone calls allowed on airplanes, if there's no harm in it? The real reason has nothing to do with the FAA; it comes from the

FCC, the Federal Communications Commission. It has nothing whatsoever to do with safety or security. When you're seven or eight miles up in the air, your phone can hit any of hundreds of cell towers, and there is supposition that this could cause a problem. As we know from 9/11, cell phones work fine from the air, but nevertheless the FCC has enacted a law making it illegal to operate a cell phone in a commercial plane that's not touching the ground. I've used mine from general aviation aircraft on a number of occasions and never had a problem either. A non-profit called RTCA is the Federal Advisory Committee for the FAA, and their report finding that cell phones pose no risk to aircraft safety is detailed in their report DO-235A, *Assessment of Radio Frequency Interference Relevant to the GNSS.* The only law that the FAA has is in support of the FCC law.

Boeing and Airbus routinely bombard their aircraft to harden them against every conceivable type of attack, physical and electronic, certainly including cell phone signals. If cell phones had the potential to endanger an aircraft, you'd be allowed to bring them on board in the same way as you bring dynamite on board. Meaning, not at all.

All other devices that you're not allowed to use during takeoff and landing (PDA's, video games, iPods, laptops), are not restricted by either the FAA or the FCC. You'll find the authority for this in RTCA document DO-233, *Portable Electronic Devices Carried on Board Aircraft.* These rules are arbitrary and are invented by the airlines, without any legal authority. It is their plane and they're within their rights to make whatever rules they want, but travelers should know that there are no laws against using these devices at any time, and that the research has been conducted and the devices have been demonstrated to be safe. Feel free to put this on the comment card next time you fly.

When you listen to the flight attendants explain the rules, it's clear that their training includes a simple mention that portable electronics and cell phones represent a danger. The trainers probably believe it and the flight attendants have no reason to question what they've been told. If you've ever wondered how your 1.5 volt LCD Palm Pilot could be so dangerous, you were on the right track. When you hear something that sounds far fetched, be skeptical.

So what's the deal? Are Men in Black secretly going from airline boardroom to boardroom, handing out corporate welfare payments in exchange for the promise to support the government's evil plan to convince us all that we're on the brink of destruction? Are there paranoid, over-cautious, or ignorant policymakers in charge at all the airlines? Personally both of those are a little too conspiracy theory for my taste, but I also think there might be a small element of truth in each.

Take the example of the terrorist train bombing in Madrid in March of 2004. The bombs were set off by cell phone calls, since cell phones are easy to get and are reliable. One reaction, which thankfully has not been put in place so far that I've seen, was the immediate proposal to jam cell phone signals anywhere that was bombable. Airports, national monuments, stadiums, train stations. Was this a logical reaction? No. If the bombers couldn't have used cell phones they would have used walkie talkies from Radio Shack. If they couldn't have used those they would have used mechanical wind-up clocks attached to detonators. If they couldn't have used those, they would have used slow burning fuses. There are a million ways to set off a bomb and no law can prevent it from happening. Any reasonable person understands this. Unfortunately, our politically correct, alarmist, liability hysterical culture demands that government do something. The culture doesn't know or care whether it's logical, or makes any difference. Our culture sleeps better knowing that Big Brother is babysitting. Satisfy the public's emotions, and you have a happy population. I guarantee you that Osama bin Laden is not rending his garments in despair over all his plans being ruined, now that Americans aren't allowed to bring a bottle of water on board a plane. It's a useless and inconvenient law, but it shows that government cares, and our culture is willing to be walked all over and curtailed in any way necessary to feel protected.

I think the cell phone ban on airplanes is just another example of this. Big Brother has cultivated and nourished a supposed danger, and leveraged it into another opportunity to babysit. Now I promise you that I'm not a conspiracy theorist, and I'm not some paranoid anti-government guy who thinks the government is out to get us. But I can't think of a better explanation for the absurd inflight cell phone ban, than the one I've just given.

15. SUV Phobia

Let's spend some time on the trendy fad of looking for villains to blame for global warming. My flavor of the week is SUV's, those evil gas guzzling, ozone destroying, unethical, politically incorrect, Nazi family soccer wagons. Only let's not do it the trendy way, let's look at the issue skeptically.

Let's start by finding some common ground, some generalizations that most people probably agree with. First, the premise that fuel efficiency in vehicles is a good thing. There are probably very few people who disagree that efficiency should always be a goal. Second, the premise that heavier cars are generally less fuel efficient, thus lighter cars are generally good things. Excess weight burns excess fuel. Cars should not be unnecessarily heavy. Third, many heavy truck-based SUV's are generally heavier and less fuel efficient than light passenger cars.

I'm going to continue with the assumption that you agree with all of the above. Based in part on these generalizations, many so-called environmentalist groups have been lobbying, often successfully, for laws against SUV's. I hope to encourage you to be skeptical of such laws. The problem with making laws based on generalizations is that the exceptions are being unfairly penalized, and some guilty offenders are getting away scott free. Any law against SUV's is a bad law, and here's why.

The vast majority of so-called SUV's are mechanically identical to conventional cars. They are given taller bodies and more upright styling, then sold as SUV's. Their weight, economy, and performance are generally similar to the cars on which they are based. Toyota's Highlander and Lexus RX series are built on Toyota Camry chassis and mechanicals. Honda CRV and Element SUV's are based on the Honda Civic. The Toyota Matrix and Pontiac Vibe are rebodied Toyota Corollas. The Hyundai Tucson, Santa Fe, and Kia Sportage SUV's are based on the Hyundai Elantra and Sonata sedans. The Acura

MDX and Honda Pilot SUV's are simply Honda Accords underneath that taller sheet metal. People don't need heavier metal or tougher mechanicals, they simply want a particular cosmetic style or a form factor that's more convenient for carrying people and cargo. And that's fine.

For example, a military Humvee, now also marketed to consumers by General Motors as the H1 Hummer, has portal axles and inboard brakes. Most people don't know what either of those are, but suffice it to say that they represent dramatic structural departures from conventional SUV's. People want to buy a big beefy military vehicle, but GM's engineers know that it's simply not a practical road car. Not wanting their customers to be disappointed, they took their existing conventional Yukon/Tahoe/Escalade vehicle, put a vaguely Humvee-like body on it, and they now sell it as the H2 Hummer. Most people wrongly assume, as GM hoped they would, that it's a second generation Humvee, new & improved, but still with military vehicle roots. Wrong on all counts, but again, most consumers don't know or really care. Not a single component is shared between the H1 and H2. Their whole design paradigms are polar opposites: one is a military truck, the other is a passenger car with a styled exterior. GM knew that people wanted to believe that they're driving a Humvee, so GM tried to license the name Hummer from the Humvee's manufacturer, defense contractor AM General; but AM General refused. GM had to buy the entire company, just to get access to the Hummer name so they could sell more H2's. It was well worth it since GM sells an H2 Hummer for about twice the price of a mechanically identical Yukon or Tahoe. And consumers now blissfully believe they're driving around in military trucks. Yet another example of why you should be skeptical of marketing labels.

People talk about cleaning up Los Angeles' smog by penalizing or banning SUV's. Did you know that a single container ship coming into Long Beach Harbor generates as much carbon emissions as 300,000 cars? Ships are not subject to emission laws. Why not? Are SUV's, most of which are mechanically and economically similar to conventional cars, really the logical targets? SUV's are hardly the cause of our carbon problems. Any road car, H2 Hummers included, is extremely environmentally friendly (as vehicles go), given all the emission laws that they comply with, especially when compared to the average car from only a decade ago.

Paris and London are two cities that have really gone agro over SUV's, fining them for entering downtown. The claim is that they're not only fuel inefficient, but they're too big to park and too dangerous. But, as we've established, the term SUV really only refers to cars with a certain cosmetic style. There are plenty of cars that are fuel inefficient that are not SUV's. There are plenty of cars that are longer than many SUV's. And there are plenty of cars that are tall or heavy and do as much crash damage as SUV's. SUV's probably appear frequently on all three lists, but targeting cars because of their styling is still the wrong path to a useful solution. Ban cars that are fuel inefficient, or ban cars that are too long to park, or ban cars with bad crash ratings. Even do all three. But you won't solve those problems by attacking the irrelevant characteristic of cosmetic styling. So why do lawmakers do it? They don't care about the facts, they care about appealing to the voters' emotions. Ban those evil SUV's, and you'll satisfy the emotions of the ignorant masses. If you're not ignorant, you shouldn't stand for it. You should demand that lawmakers pay attention to the facts. (You might also mind your own damn business and stop trying to legislate what other peoples' priorities should be, but that's another subject for another time.)

Here's another wrinkle for you. Hybrids such as the Toyota Prius and Honda Insight have really pushed the bar higher on efficiency and economy. Thus, there is now a general perception that hybrids get better mileage. Generally true, but again, there are exceptions. The Lexus RX hybrid SUV uses exactly the same V6 engine as its non-hybrid counterpart, and thus posts similar mileage numbers. I drove both vehicles prior to their release in a consumer test. The hybrid system in this case simply adds additional power for acceleration. The improved mileage that you might expect from the hybrid system is canceled out by the additional weight of the battery and motor, particularly on the highway. The Lexus GS is an example of the same philosophy applied to a high-end luxury sedan. In addition, many high-end sports car manufacturers are testing hybrid prototypes for the electric engine's ability to add acceleration off the line. In summary, a hybrid system does not always mean improved economy or cleaner emissions. You should pay attention to the actual numbers that a vehicle posts, not to its label, be it "hybrid" or "SUV".

Here's the first example that pops into my head: my 2004 Audi S4, a 4 door sedan, gets 15 miles per gallon, which is worse than the 16 miles per gallon of my wife's 2006 Toyota 4Runner with the largest V8 engine. Which do you hear so-called environmentalists protesting: common sedans, or SUV's? They're smart: Protesting sedans will strike no nerves, but it's easy to terrify the public with alarmist warnings about those evil SUV's. And I think that this perfectly summarizes the fact that anti-SUV protests and legislation are not only counterproductive, they are factually wrong. When you hear marketing buzzwords and labels instead of valid test data, be skeptical.

16. THE REAL PHILADELPHIA EXPERIMENT

In this chapter we're going to pull a giant switch on the wall and activate powerful generators, which will create a mysterious force field around us and cause us all to disappear. For we're going to discuss the perennially silly "Philadelphia Experiment."

The story goes that in October of 1943, at the Philadelphia Naval Shipyard, an experiment was conducted aboard a US Navy Cannon-class destroyer escort called the USS *Eldridge*, number DE-173. The experiment involved the creation of a force field which rendered the ship invisible both to the eye and to radar. The experiment was witnessed by hundreds, possibly thousands, of sailors both ashore and on other ships nearby. Unfortunately, there were severe side effects to the crew on board ship. Some were found materialized inside the metal of the ship, others were never seen again, and still others were driven insane or plagued for years by mysterious cases of phasing in and out of existence. In typical Navy fashion, everything has been denied.

None of the above ever actually happened. What follows, *did* actually happen.

The secret of the mysterious experiment and its terrible aftermath was kept for 12 years, until one day when Morris Jessup, the author of a book about UFO's, was unexpectedly summoned to the US Navy's Office of Naval Research. The ONR had received, in the mail, a copy of Jessup's UFO book, filled with handwritten annotations. The notes were all in the same handwriting, but in three characters, in three colors of ink. The notes revealed all the details of the Philadelphia Experiment, and the ONR wanted to know if Jessup knew anything about who might have written these notes and sent the book to them.

Jessup recognized the crazy handwriting immediately. He had received a series of letters from a man calling himself Carlos Allende, who claimed to have witnessed the experiment from aboard a merchant marine ship nearby, the SS *Andrew Furuseth*. Jessup had dismissed Allende as a crackpot — among Allende's claims was that Albert Einstein had personally spent several weeks mentoring him on subjects such as invisibility and faster-than-light travel. Several copies of the annotated book were produced, and the rest is history: The Philadelphia Experiment story became part of the fabric of pop culture.

Robert Goerman, a researcher of the paranormal and a friend of mine (by email), noticed that the return address of Allende's original mailing to the ONR came from his own home town, New Kensington, Pennsylvania. He had spent some time putting together the history of the Philadelphia Experiment story and in 1979, found to his great surprise that Carlos Allende, whose real name was Carl Allen, was the son of a close family friend. Over time, Goerman filled in the blanks and presented the complete case in magazine articles and on television shows such as *History's Mysteries, Unsolved Mysteries,* and *The Unexplained* — three show titles which by themselves lend far too much credence to Carl Allen's stories. Carl Allen was something of the dark horse of the family, a creative and imaginative loner, notorious for annotating anything and everything in the house, and sending bizarre writings and claims to everyone in the family for any occasion. Goerman also assembled all of the facts of the USS *Eldridge:* little things, like the ship was nowhere near Philadelphia when the experiment happened, the ship had detailed corroborated records for the normal duties that it was in fact performing at the time, and that nobody who had ever been a crewman on board the ship knew anything about any experiment. The *Eldridge* had been launched only two months before the experiment, and you'd think that if it was hosting Albert Einstein and the most amazing experiment in history, somebody would have known about it. Without exception, every single fact that Carl Allen presented as evidence of the Philadelphia Experiment has been easily proven to be a complete fabrication.

And therein lies the problem. The trouble with discussing government conspiracies is that the believers generally refuse to accept the factual evidence, because it becomes part of the conspiracy. It's a bit like having a debate about Creationism: believers simply say "God did it" and it's a matter of faith, not of evidence or fact. If the government is trying to cover something up, every falsified claim becomes evidence for the conspiracy. In summary, there is no amount of evidence that can be compiled that will be accepted by a conspiracy theorist. In the conspiracy theorist's mind, evidence *against* is actually evidence *for.*

Among this schizophrenic evidence is the US Navy's reply to the Philadelphia Experiment story, which is available online at http://www.history.navy.mil/faqs/faq21-1.htm. As you might surmise, it simply says that they have no idea what this guy could be talking about, here's where both ships were and what they were doing at the time, and here are the statements from their officers. It's an entertaining report, but predictable.

Interestingly, there are some question marks left standing about the Philadelphia Experiment. As you might expect, two or three veterans of the *Eldridge* and the *USS Engstrom,* which was once moored alongside the *Eldridge* in 1943, have claimed to be a guy whom Allende saw disappear in a bar or who have found themselves transported across dimensions. It should be noted that such claims were made well after the Philadelphia Experiment became a pop-culture phenomenon, and in more than one case it turned out that these guys were not actually veterans at all. A more interesting question comes from the very genesis of the story, when the Navy ONR first summoned Morris Jessup to their office to talk about the strangely annotated book. If there was nothing in those annotations of genuine interest to them, why did they call?

Who knows. Goerman suggested that these ONR officers did this on their own time out of personal intrigue, or that some of the allusions in the annotations were similar to actual Navy research and the ONR was simply being responsible at following every lead. Conspiracy theorists maintain that this proves Allen's annotations must have been true. Whatever the truth, I happen to love that this story has this annoying little footnote that can't be easily explained away. I love a good mystery and I remember a streak of disappointment when Goerman's debunking of the case was laid out before me — like most people I'd done no thorough research of my own and only knew what the average moviegoer knows about the Philadelphia Experiment. Intellectually I know the story is silly, but I still love that little loose end. Carl Allen may have been a prankster, but I thank him for making the century just that little bit more interesting.

17. INTERNET PARANOIA

Next we're going to take a skeptical look at computer security. How real are the threats we are warned about all the time? Do these supposed threats pose any actual danger, and if so, what kind? What steps do we really need to take? This stuff is my business, so I know of what I speak.

In the early days of web browsing, innovative programmers created cookies as a way to store session variables on the visitor's computer. When Bob visits Amazon and tells the server that his name is Bob Smith, Amazon writes "Bob Smith" on a nametag and sticks it on Bob's shirt. This is a cookie. It permits Amazon to accumulate a shopping cart full of merchandise for Bob, and to know where to put each new item without asking Bob to identify himself each time he brings something else to the register. The server can say "Ah, you're Bob Smith, good to see you," and it knows which shopping cart to put Bob's new DVD into. Without cookies, it would be necessary for Bob to log in each time he adds something to his cart. Not only is this inconvenient for Bob, it requires substantially more server resources. Server resources are not infinite. The more efficiently a web server can run, the faster it can serve Bob. And, when Bob's name tag is prominently plastered to the front of his shirt in the form of a cookie, there's less chance for Bob to be misidentified and be sent the wrong merchandise. Cookies are good for everyone.

Skeptoid.com even uses cookies, albeit in a simpler way. When you submit a comment on a podcast episode, the name and hometown that you enter into the comment form are saved on your computer as cookies. This allows Skeptoid.com to automatically fill in these fields for you, the next time you want to submit a comment. Saves you a few dozen keystrokes, and makes your entries more consistent. Just a minor convenience feature. Minor, but still a good thing.

Unfortunately, in the early days of Microsoft Internet Explorer, some genius in Redmond decided that Internet Explorer might seem superior to Netscape — its main rival at the time, if you remember — if it would raise caution flags and warn you about terrifying security risks with scary dialog boxes. Internet Explorer eventually became the only significant web browser and a whole generation of web surfers was raised with the belief that cookies were a security risk to be feared and blocked at all costs. The idea is that an unscrupulous individual might sneak into Bob Smith's office, look in the cookies folder on his computer, and learn that "Bob Smith" is the name he used to identify himself to Amazon.

Cookies are just plain vanilla text files. They contain no program logic or encryption. The good thing about this is they can't contain program code like viruses or trojan horses. The bad thing about this is they contain human readable plain text, so that anyone with access to your computer can read them. Since nearly every other program and data file on your computer uses human readable plain text, is this truly such an egregious security risk? Programmers and software engineers know that it's not, but marketing people never let the truth stand in the way of a sale. If they can convince you that your computer's normal operation constitutes a risk that can be mitigated by purchasing their software, they know they've got lots of sales.

Some people think that cookies can be used to steal credit card numbers or other information from your computer. Not only is there no mechanism by which this could work, it's illogical. The web server is what writes the cookie to your machine, and obviously it can't write anything it doesn't already know.

Referrer codes are another normal function that's being marketed as a security risk. Whenever a web browser visits a server, it sends a referrer code. This is the URL of the web page from which the browser came. This is part of the http specification and is a normal function, it's not the nefarious evil plan of some hacker. Let's say our friend Bob is reading the news on CNN.com and sees an ad for a plasma television from Amazon. Bob's in the market for a good plasma, so he clicks the ad. Amazon's web server receives a referrer code from Bob's browser that tells it Bob linked from CNN.com. Amazon may use the referrer codes to analyze which of their advertisements are most effective, an analysis that's essential to good advertising. If Bob buys something, CNN or some third party may be entitled to a sales commission for referring the business, which Amazon is happy to pay since they're happy to have Bob's business. Amazon may even see where Bob came from and offer him the special CNN discount. The referrer code is great for Amazon. At best it's great for Bob, at worst it's no skin off Bob's nose. Referrer codes are also used for many other useful things on the web.

As you might expect, the security software vendors market referrer codes as a threat too. Their best explanation is that it's none of Amazon's business where you came from. That's true, in a strictly Libertarian sense, but in a practical sense, it's really helpful for them to know. Many services such as Amazon can ' better customize their offerings when they know where their visitors are coming from. A technology called Collaborative Filtering allows Amazon to say "Visitors from CNN prefer the new Barack Obama music video." If you use security software to block your browser from sending referrer codes, the best you'll get is a more generic Internet experience. The worst you'll get is that some web services won't work at all.

Viruses are a genuine pain in the ass. If you're running Windows and you use the Internet at all, your computer will probably download at least a dozen new ones a day. There are numerous ways that viruses can be embedded in web pages, in advertisements on web pages, in media presentations, even in some older graphic images. Big money is made by some of the more cleverly designed viruses (or trojan horses, whatever term you prefer). Sometimes all they do is install fake cookies on your computer to tell Amazon that some guy in Indonesia is entitled to a sales commission next time you buy something. Sometimes they make your computer part of a distributed illegal file sharing system. Sometimes they turn your computer into a spam server. Sometimes they install browser toolbars that lead you to more virus pages when you use them. Sometimes they install software that displays popup advertisements. Fortunately, free software such Spybot Search & Destroy or Ad-Aware can genuinely eliminate all, or almost all, of these threats. So what's the Skeptoid angle on these?

The reason people develop viruses and trojan horses is to make money, through advertising or sales commission programs. Really all of the threats described above ultimately lead to money. However, in point of fact, it's no skin off your nose. Sure, some guy in Indonesia will get a sales commission that he didn't deserve, but it didn't come out of your pocket. And it benefits nobody to write a virus that erases your computer or causes some problem. Such viruses do exist, mainly in the early days of the Internet, but since they're not profitable, they've gone almost completely out of fashion. People want your business and they want your money, there's no profit in erasing your hard drive. Invariably, when I make this rant, I get the comment "My aunt had her computer erased by a virus," or "I have a virus that makes my screen go black and deletes my address book." The truth is that software conflicts and system crashes are more likely responsible for these problems. If you own a computer, problems are the price of admission, and you will eventually lose data. But there are a thousand normal operating system problems that will be the most probable culprit. It makes no sense for someone to write a virus that does these things, when they can just as easily write a virus that earns them money. Just as in nature: the viruses that thrive are those that don't kill their hosts.

Use Spybot Search & Destroy and Ad-Aware to scrape this crap off your machine and keep it running lean and mean, but don't buy the expensive commercial software that does no better and that makes claims designed to take advantage of customers with minimal technical knowledge. Or just use a Mac like I do, since none of this crap runs on Unix. And, don't bury your thoughts too deeply in cyberspace. Remember you'll always be most vulnerable to what is statistically by far the biggest threat: that your computer will simply be stolen.

18. The "New" Bill of Rights

The first ten amendments to the United States Constitution are called the Bill of Rights. It was adopted in 1791, two years after the Constitution went into affect. Some have said that the Bill of Rights represents one of mankind's greatest leaps forward, establishing a new and previously unheard of standard for personal and national liberty. However, in recent decades, it's begun to show its age, and is no longer relevant to the lives of modern Americans. It no longer represents our politically correct culture. So, I hereby propose this amended Bill of Rights to better reflect what Americans truly want.

First Amendment – *Freedom of speech*
You have the right to never be exposed to speech which might possibly offend someone somewhere. The government shall maintain a Federal Communications Commission to thoroughly censor all broadcast media, and impose strict fines on any and all offensive content.

Second Amendment – *Right of the people to keep and bear arms*
You have the right to be guaranteed that no law abiding citizens living near you may ever be armed with dangerous weapons.

Third Amendment – *Protection from quartering of troops*
No soldier shall, in time of peace, be quartered in any house, without the consent of the Owner, unless that house is in some foreign country.

Fourth Amendment – *Protection from unreasonable search and seizure*
The right of the people to be secure shall be protected by frequent searches and seizures upon persons of a different race. The unreasonable cruelty of a warrant shall not be imposed.

Fifth Amendment – *Due process, double jeopardy, self-incrimination, private property*

No person shall be held to answer for any crime, unless adequate due process be applied, and applied, and applied, and applied, and applied. Private property shall not be taken for public use, except to create a Wal-Mart.

Sixth Amendment – *Trial by jury and other rights of the accused*

In all criminal prosecutions, the accused shall enjoy the right to a speedy and public trial, and to be released from all responsibility for that crime if enough Hollywood celebrities feel that he has turned over a new leaf.

Seventh Amendment – *Civil trial by jury*

In any and every dispute in business, family, sports, or entertainment, where the value in controversy shall exceed twenty dollars, a court of the United States shall always be called upon to settle all matters through lengthy and expensive litigation.

Eighth Amendment – *Prohibition of cruel and unusual punishment*

Cruel and unusual punishment, such as mishandling your Koran or making you perform a human pyramid, shall never be inflicted, except in fraternity houses.

Ninth Amendment – *Protection of rights not specifically enumerated in the Bill of Rights*

The enumeration in the Constitution of certain rights, shall not be construed to mean that people have any other rights. In fact you are guaranteed that people you don't like, or who are of a different ethnic background than you, shall have no implied rights at all.

Tenth Amendment – *Powers of states and people*

Neither the states nor the people shall ever infringe on your rights to have the federal government force everyone, in every state, to adopt your personal opinions.

These proposed amendments are humbly submitted by the majority of the American public, excepting only those who prefer that the Bill of Rights be replaced by the Ten Commandments. For their speedy adoption will this petitioner ever pray.

19. ORGANIC FOOD MYTHS

We're now going to put on our tie dyed shirt, grow our hair long and dirty, claim hatred for science and corporate America, then walk into the most expensive specialty supermarket in town and purchase one of the most overpriced products on the market today: Organic food.

Organic food is a conventional food crop (genetically exactly the same plant variety as the regular version) but grown according to a different set of standards. In this sense, organic food is really the same thing as kosher food. The food itself is identical, but it's prepared in such a way to conform to different philosophical standards. Just as kosher standards are defined by rabbinical authorities, the USDA's National Organic Program sets the requirements for foods to bear a "certified organic" label. Basically it forbids the use of modern synthetic fertilizers and pesticides in favor of organic equivalents, and for animals it requires that they have not been kept healthy through the use of antibiotics. There are other rules too, and the basic goal is to require the use of only natural products throughout the growth, preparation, and preservation stages.

Organic food is more expensive than conventional food, due not only to its lower crop yields and more expensive organic fertilizers and pesticides in larger quantities, but mainly because it's such a big fad right now and is in such high demand.

Why is that? Is organic food healthier? Does it make an important political statement? The usual arguments boil down to three: that it benefits small farmers rather than big evil companies; that it's somehow healthier to eat; and that the cultivation method is better for the environment. Rather than accepting these emotionally satisfying benefits at face value, let's instead take a skeptical look and see what the data actually show. Let's take these three claimed benefits one at a time.

Buying organic food benefits small farmers, and represents a blow to the big food corporations.

All right, let's take for granted the position that major food producers deserve to be struck with a blow. I'm sure the starving millions in Africa appreciate the sentiment.

Make no mistake, organic food is big, big business. The days when the organic produce section of the supermarket represented the product of a small local farmer are long gone. California alone produces over $600 million in organic produce, most of it coming from just five farms, who are also the same producers of most non-organic food in the state. 70 percent of all organic milk is controlled by just one major milk producer.

Five or ten years ago, when the major food producers saw that organic food was coming into vogue, what do you think they did? They smelled higher prices charged for less product, and started producing organic crops. Nearly all organic crops in the United States are either grown, distributed, or sold by exactly the same companies who produce conventional crops. They don't care which one you buy. You're not striking a blow at anyone, except at your own pocketbook.

Trader Joe's is a supermarket chain specializing in organic, vegetarian, and alternative foods with hundreds of locations throughout the United States, centered in organic-happy Southern California. Shoppers appreciate its image of healthful food in a small-business family atmosphere. Really? In 2005 alone, Trader Joe's racked up sales estimated at $4.5 billion. The company is owned by a family trust set up by German billionaire Theo Albrecht, ranked the 22nd richest man in the world by Forbes in 2004. He's the co-founder and CEO of German multi-national ALDI, with global revenue in grocery sales at $37 billion. According to Business Week, the decade of the 1990's saw Trader Joe's increase its profits by 1000%. Trader Joe's also compensates its employees aggressively, with starting salaries for supervisors at $40,000. They hire only non-union workers. Now, to any capitalist or business-minded person, there's nothing wrong with any of that (unless you're pro-union or anti-big business). It's a great company, and very successful.

Trader Joe's customers are willing to pay their premium prices to get that healthful image. But they should not kid themselves that they're striking a blow at big business and supporting the little guy.

I'm not exactly sure why anticorporatism wound up on the organic food agenda, since it's so counterintuitive. The irony is that the organic food companies supply a smaller amount of food per acre planted, and enjoy dramatically higher profits, which is why anticorporatists hate corporations in the first place.

Organic foods are healthier to eat.

Did you ever wonder why Chinese drink only hot tea? They boil it to kill the bacteria. Most local Chinese farming uses organic methods, in that the only fertilizers used are human and animal waste: Without being boiled, it's basically a nice cup of *E. coli*. In the case of China and other poor Asian nations, the reason for organic farming has less to do with ideology and more to do with lack of access to modern farming technology.

The National Review reports that Americans believe organic food is healthier by a 2-1 margin, despite the lack of any evidence supporting this. When you take the exact same strain of a plant and grow it in two different ways, its chemical and genetic makeup remain the same. One may be larger than the other if one growing method was more efficient, but its fundamental makeup and biochemical content is defined by its genes, not by the way it was grown. Consumer Reports found no consistent difference in appearance, flavor, or texture. A blanket statement like "organic cultivation results in a crop with superior nutritional value" has no logical or factual basis.

Some supporters of organic growing claim that the danger of non-organic food lies in the residues of chemical pesticides. This claim is even more ridiculous: Since the organic pesticides and fungicides are less efficient than their modern synthetic counterparts, up to seven times as much of it must be used. Organic pesticides include rotenone, which has been shown to cause the symptoms of Parkinson's Disease and is a natural poison used in hunting by some native tribes; pyrethrum, which is carcinogenic; sabadilla, which is highly toxic to honeybees; and fermented urine, which I don't want on my food whether it causes any diseases or not. Supporters of organics claim that the much larger amounts of chemicals they use is OK because those chemicals are all-natural. But just because something is natural doesn't mean that it's safe or healthy — consider the examples of hemlock, mercury, lead, toadstools, box jellyfish neurotoxin, asbestos — not to mention a nearly infinite number of toxic bacteria and viruses *(E. coli,* salmonella, bubonic plague, smallpox). When you hear any product claim to be healthy because its ingredients are all natural, be skeptical. By no definition can "all natural" mean that a product is healthful.

Consider the logical absurdity proposed by those who claim conventional growers produce less healthful food. To the organically minded, conventional growers are evil greedy corporations interested only in their profit margin. What's the best way to improve the profit margin? To buy less pesticides and fertilizer. This means they must use far more advanced and efficient products. The idea that pesticides leave dangerous residues is many decades out of date. Food production is among the most regulated and scrutinized of processes, and today's synthetic pesticides and fertilizers are completely biodegradable. They're supported by decades of studies that demonstrate their total safety.

In the United States, 2006 brought two major outbreaks of *E. coli,* both resulting in deaths and numerous illnesses, ultimately traced to organically grown spinach and lettuce. According to the Center for Global Food Issues, organic foods make up about 1% of all the food sold in the United States, but it accounts for 8% of *E. coli* cases.

Organic growing methods are better for the environment.

Organic methods require about twice the acreage to produce the same crop, thus directly resulting in the destruction of undeveloped land. During a recent Girl Scout field trip to Tanaka Farms in Irvine, California, one of the owners told us his dirty little secret that contradicts what you'll find on his web site. Market conditions compelled them to switch to organic a few years ago, and he absolutely hates it. The per-acre yield has been slashed. Organic farming produces less food, and requires more acreage.

Many so-called environmentalists generally favor organic farming, at the same time that they protest deforestation to make room for more agriculture. How do they reconcile these directly conflicting views? If you want to feed a growing population, you *cannot do* both, and soon won't be able to do either. If you support rainforest preservation, logically you should oppose organic farming, particularly in the developing world. On the other hand, if you demand organic soybeans, then you should have the courage to stand up and say that you don't care whether black and brown people around the world have enough to eat or not.

I'm not making this stuff up. For every dreadlocked white kid beating a bongo drum in favor of organics, there is a Ph.D. agriculturist warning about its short sightedness and urging efficient modern agriculture to feed our growing population. Personally I like forests and natural areas, so I favor using the farmlands that we *already* have as efficiently as possible. This benefits everyone. I say we dump the useless paranormal objections to foods freighted with evil corporate hate energy, and instead use our brains to our advantage for once. When we find a better way to grow the same crop faster, stronger, healthier, and on less acreage, let's do it. We all benefit.

20. THE REAL AMITYVILLE HORROR

In the small town of Amityville on New York's Long Island, on a dark evening in 1974, 23 year old Ronald "Butch" DeFeo burst into a bar and declared that his entire family had just been shot. Police discovered six bodies in the DeFeo home at 112 Ocean Avenue, and what's more, the subsequent investigation revealed that Butch DeFeo had himself killed them all: both his parents, and his four younger siblings, with a Marlin rifle. Despite DeFeo's claim that strange voices in his head compelled him to commit the murders, he was convicted of all six murders and remains imprisoned to this day.

Just over a year after the murders, the home was purchased by newlyweds George and Kathy Lutz, who moved in with their three children. The house was sold furnished so all of the DeFeo's furniture was still there, just as it had been on the night of the murders. George Lutz had heard of the murders, so just to be on the safe side, they called a priest whom Kathy knew, to bless the house. The trouble began when the priest was driven out of the house by an angry disembodied voice, and received stigmatic blisters on his skin. The family daughter reported a friendly pig named Jodie, who later began making appearances to the rest of the family through windows. A sculpted lion came to life and walked around the house, and even bit George Lutz. The apparition of a demonic boy appeared and was photographed, which you can find online. Angry red eyes looked into house at night, and left cloven footprints in the snow. George Lutz woke up in a sweat every night at the same hour the DeFeos were murdered. Stephen Kaplan, a local parapsychologist, was called in to investigate. Powerful forces caused doors to explode off their hinges. Kathy developed strange red marks on her chest and levitated two feet off her bed, and George saw her transform into a hideous old hag. Green slime oozed from the walls of the house, and a crucifix on the wall constantly rotated itself upside down. And, in one final night of terror that the Lutzes have never even been able to describe, the family was driven out of the house, never to return. Their stay had lasted only 28 days.

The events are not surprising, since a few hundred years before the DeFeos were murdered, the local Shinnecock Indians used the same property as a sort of insane asylum for their sick and dying. Negative demonic energy was nothing new to the Amityville Horror house.

So what happened next?

George Lutz, whose business was failing (ostensibly as a result of the distraction of the haunting), hoped to find a silver lining and called up the publisher Prentice-Hall. *The Exorcist* had come out only two years before and had been wildly successful, putting things like demons and abused priests firmly in the public consciousness, so Prentice-Hall was keen to capitalize on the Lutzes' experience. The publisher engaged author Jay Anson to write the book *The Amityville Horror,* and the rest is history. The book and subsequent *nine* motion pictures were highly successful in the aggregate, though most critics agree that the movies were all pretty stupid.

Where it started to get murky was a meeting that George Lutz had during his 28 days in the house. The man he met with was William Weber, who was no other than Butch DeFeo's defense attorney. Who initiated the meeting is not clear. According to William Weber's admission in later years, what transpired in that meeting was an agreement that served both men's interests. The story of the haunting was concocted, based in part upon elements from *The Exorcist.* George Lutz stood to gain from the potential commerciality of a ghost story based upon the DeFeo murders, and Weber would have a new defense for his client: Demons, as evidenced by the Lutzes' experience, caused Butch DeFeo to murder his family, at least in Butch's own mind.

Prior to the publication of the book, Lutz and Weber set out to publicize the haunting to the best of their ability. The most notable success they had was a television crew from Channel 5 in New York who brought a pair of psychics, Ed and Lorraine Warren, who reported on camera that the house was plagued with malevolent spirits. Other psychics also visited the house on different occasions, and by the time the book came out, the groundwork for a bestseller was well laid. To this day, the Warrens maintain that the Amityville Horror was a genuine haunting, and they describe their experience on their website warrens.net.

Once enough money had been made by the book, the lawsuits began. Lutz and Weber sued each other and just about everyone else under the sun, with claims such as breach of contract, misappropriation of names, and mental distress. The judge eventually threw everything out of court, along with a stern lecture about the book being a work of fiction based in large part upon Weber's suggestions and the popularity of *The Exorcist*.

Let's back up a moment to Stephen Kaplan, the parapsychologist whom the Lutzes called in during their brief stay in the house, per Lutz and Weber's marketing plan. As it turned out, Kaplan didn't buy the Lutzes' story, and concluded while he was there that he was being hoaxed. Later, when Anson's book became so popular, Kaplan became concerned that the Lutzes' story, which he considered bogus, would give paranormal research a bad name, so he wrote his own book called *The Amityville Horror Conspiracy* in which he laid out more than 100 factual inconsistencies.

Among these inconsistencies was that the Shinnecock Indians, or any other native tribes, never lived anywhere near present-day Amityville. The nearest Shinnecock settlement was 70 miles away, even according to the Shinnecocks themselves, and you can find their web site at shinnecocknation.com.

Father Pecoraro, the priest who tried to bless the house when the Lutzes moved in but was allegedly attacked, reports that nothing unusual happened during his visit and no attack or evil threat of any kind took place, although he did express a concern about evil spirits to George Lutz based on the house's history. Some reports, including one affidavit by Pecoraro himself, state that he never visited at all but only spoke to the Lutzes over the phone. As a result, author Jay Anson created a new priest for the book, Father Mancuso. The only priest who ever got blisters and a ghostly warning was a fictitious character.

There were many other inconsistencies. No doors or windows in the house were found to have ever been ripped off their hinges; all were found undamaged and securely mounted with their original hardware. The local police department records indicate no calls or visits to the property during the 28 days, despite a number of such events in the book. The same goes for disturbances affecting the neighbors. No snow fell during the period, which strains the cloven hoof prints in the snow. In short the book was full of episodes that created physical evidence, but none of that alleged evidence has ever withstood any scrutiny. Indeed, everything that was falsifiable *was* easily falsified by Kaplan and numerous other investigators. It should be noted that the Warrens, who enjoyed their best public exposure during their televised visit to the house, consider Kaplan's book to be false and to be simply his own attempt at self-serving publicity.

By now you know what my recommendation is when you encounter claims that are far fetched or that violate physical laws: Be skeptical. In the case of the Amityville Horror, plenty of evidence exists to indicate that none of the events in the book happened, and the only evidence that anything *did* happen are anecdotal personal accounts by parties with clearly vested commercial interests. If you want to read the book — and most readers report that it is a great scary story — enjoy it for a work of fiction that launched one of pop culture's most engaging and long-lived ghost stories.

21. LIVING STONES OF DEATH VALLEY

If you're familiar with the American Southwest or even if you're a fan of the paranormal, you've heard of the mysterious stones that move across the surface of a dry lakebed in Death Valley called Racetrack Playa. Hundreds and hundreds of rocks, scattered about the surface of this several square mile mudflat, have left trails behind them where they've moved across the surface. Nobody has ever seen one move, despite many studies. I came as close as anyone could.

Proposed explanations run the gamut from natural to paranormal to alien. Strange magnetic forces, psychic energy, alien spacecraft, teenage pranksters, and even transdimensional vortices have all been proposed. The leading scientific hypothesis is that the rocks are moved by high winds, on rare occasions when the playa is wet enough to be extremely slippery, and conditions are just right. I've always had trouble with this explanation. I used to play in mud flats as a kid, and when a rock is glued onto that surface it's pretty damn hard to move. The rocks at Racetrack Playa are quite streamlined, and it's hard to imagine any wind strong enough to break their bond with the surface and shove them along deep enough to leave those trails. The real cause of the moving rocks, it turns out, carries a lot more punch than wind, and requires conditions that are not oddball and that are easily observable.

In the early spring of 2002, I made one of my many trips to Racetrack Playa with two friends, Dan Bocek and John Countryman. The surrounding mountains were still covered with snow, and the playa itself was firm but had a large lake covering about a fifth of its surface, perhaps an inch or two deep at its edges, concentrated at the playa's south end where it's lowest. We ventured out, armed with cameras, shortly before sunrise. The temperature was just above freezing. The wind, from the south, was quite stiff and very cold. When we reached the lake, we found to our great surprise that the entire lake was moving with the wind, at a speed we estimated at about one half of a mile per hour. The sun was on the lake by now and we could see a few very thin ice sheets that were now dissolving back into water. This whole procession was washing past many of the famous rocks. It's easy to imagine that if it were only few degrees colder when we were there — as it probably had been a couple of hours earlier — the whole surface would be great sheets of thin ice. Solid ice, moving with the surface of the lake and with the inertia of a whole surrounding ice sheet, would have no trouble pushing a rock along the slick muddy floor. Certainly a lot more horsepower than wind alone, as has been proposed. The wind was gusty and moved around some, and since the surface is not perfectly flat and with rocks and various obstructions, the water didn't flow straight; rather it swapped around as it moved generally forward. Ice sheets driven by the water would move in the same way, accounting for the turns and curves found in many of the rock trails.

But don't take my word for any of this. I told you we had cameras, and I captured the event on video. Go to skeptoid.com and click on videos, and you'll see "Living Stones of Death Valley." It's well worth two minutes of your time.

That nobody has ever seen the rocks move is easy to believe. When there's water on the surface of the playa, you're not allowed to go out there — and indeed, you probably wouldn't want to. Thus there's nobody around when the ice sheets drag the rocks.

We missed the actual event, probably by a couple of weeks, so we didn't get the first real video of a moving rock. No doubt someone soon will. But we did see and document all the forces at play, and I think our explanation is far more plausible than any previous hypotheses.

22. SKEPTICISM AND FLIGHT 93

In this chapter we're going to talk about something that's perhaps still a little too near & dear to the hearts of some: Flight 93, the fourth aircraft on September 11 that crashed in Pennsylvania on its way to a target, taking the lives of all 44 people on board, including the 4 hijackers. I think it's appropriate that subject be discussed only after acknowledging that it was first and foremost a human tragedy, in which a lot of valuable family people were lost, and that we hold their memory in great respect.

There are two basic theories about Flight 93. The first, which is the government's official version, is that the plane crashed. The competing theory, favored by conspiracy theorists because it's the one denied by the government, is that Flight 93 was shot down by our own fighter planes.

Many discussions of Flight 93 that purport to be skeptical either choose one side or the other, and argue in favor of it, claiming that the competing theory is implausible, and citing all sorts of evidence in favor of whichever version they support. Well, that's not skepticism. Trying to justify a preconceived notion is simply spreading propaganda. Skepticism means to follow a critical thought process, examine all of the evidence, and arrive at a supported conclusion. I like Flight 93 as a skeptical topic, because it reminds us of what Dr. Shermer says: Skepticism is not a position, it's a process.

Flight 93 is an interesting case because the version of events favored by conspiracy theorists is, for once, not wholly implausible. We know for a fact that fighter jets were scrambled and that some were on an intercept course with Flight 93, and we know for a fact that we were prepared to shoot down any passenger jets that we had to on that day.

The National Transportation Safety Board has only the following quote about the cause of Flight 93: "The Safety Board did not determine the probable cause and does not plan to issue a report or open a public docket. The terrorist attacks of September 11, 2001 are under the jurisdiction of the Federal Bureau of Investigation. The Safety Board provided requested technical assistance to the FBI, and any material generated by the NTSB is under the control of the FBI." And, of course, whatever the FBI has determined is not publicly available, and certainly will not be at least until after their investigation is finished, and who knows when that will be if ever. So, in the lack of an authoritative explanation of the exact cause of Flight 93's crash, we can only do our best to study the available information ourselves. Let's look at four debated points.

1. The mysterious "white jet" circling the crash site.

There are reports of an unmarked white business jet circling the crash site. The official version is that controllers asked a nearby Falcon 20 business jet to descend and provide coordinates of the crash. This is corroborated by the pilot of the Falcon 20. Conspiracy theorists concede that fighter jets are not white, but they point to numerous examples of white business jets flown by the military and other government agencies, such as Customs, which they say could have been quickly armed with missiles. They also cite some reports of controllers saying that no other aircraft were in the vicinity, and that the eyewitness evidence of a white jet indicates that those controllers must be part of the conspiracy. The white jet tells us nothing about the cause of the crash.

2. Debris was found up to 8 miles away, fluttering down from the sky.

Conspiracy theorists say that this disproves the official version that has Flight 93 crash intact. However, the FBI has not released their official cause of the crash. There are at least two plausible explanations for this that don't require the plane being shot down. First, we know that the terrorists claimed to have a bomb on board. A bomb certainly could have blown a hole in the plane, releasing debris, and causing the subsequent crash. Second, the aircraft could have broken up in mid-air from aerodynamic stresses as it exceeded its maximum design speed (called the V_{NE}). The FBI has said nothing about the cause of the crash; they have not claimed definitively that it was a controlled flight into terrain by the terrorist pilot or struggling passengers. Without access to the FBI analysis, the existence of secondary debris fields tell us nothing about the cause of the crash.

3. There is an alleged 3 minute discrepancy in the times.

The published cockpit transcript ends at 10:03 with the voices of the terrorists chanting "Allah is the greatest." The NTSB analysis of the flight data recorder, infrared satellite imagery, and some air traffic controllers agree with the 10:03 crash time. Cleveland air traffic control and some seismologists put the time of the crash at 10:06. My own research was not able to find what the clocks of cockpit voice recorders are synchronized to, if anything. However, if a bomb celebrating Allah's greatness went off at 10:03, or if the aircraft suffered structural failure at 10:03, the voice recorder would be no less likely to be stopped as it would by a missile strike. The alleged missing 3 minutes tells us nothing about the cause of the crash.

4. Covering up a shootdown.

Think of all the people who must be involved with keeping track of air to air missiles. Obviously the pilot and any other pilots with him, the air boss and all the officers in the situation room, and anyone keeping an eye on the situation with radar, would know if a missile had been fired. This includes civilian controllers as well as military controllers, and anyone standing by the radar screen or at the local civilian airport's control tower water cooler talking to their girlfriend on the cellphone saying "You won't believe what just happened." September 11 was not a day when little attention was being paid to the radar screens. The airmen at the base who are responsible for loading and unloading missiles from the aircraft would know that a missile had been fired, as would their chain of command. The people who view and archive the electronic and video logs of the flight would know. Then you have the people who inventory and store the missiles - they'd know if ten went out and only nine came back. Military and civilian auditors verify these counts. Potentially thousands of people on the ground would have been in a position to see a missile being fired. Hundreds of people were on the ground at the crash site picking up wreckage, possibly including missile fragments, cataloging it, identifying it, and storing it. Let's say you disagree with me that any large number of people might be able to know that a missile had been fired. I ask you, what then is the smallest number? Fifty people at the air force base and through the chain of command? Forty? Nobody on the ground at all, or in the NTSB? That's hard for me to believe, but it's harder still to believe that even such a large number of people as that could be adequately paid off with nobody at any bank knowing it, or could be threatened by mysterious Men In Black, without a single whistle blower — especially when you consider how broadly unpopular the war on terror has become.

For my money, the official version of the incidents is consistent with my own knowledge of aviation and all sounds plausible. I also can't get past what, to me, is the implausibility of covering up a shootdown. Your own mileage may vary. But regardless of your own conclusion, better that you look at the situation with skepticism rather than with a preconceived notion, and don't base your judgment on politics or emotion, as so many people do.

There's one school of thought that says it doesn't matter how Flight 93 ended. The terrorists killed everyone on board, regardless of the details. Ultimately the terrorists are to blame, no matter the cause of the crash. Then there's the viewpoint that whether the government lied has everything to do with it: that if we can't trust our own government, how can we ever feel truly safe under its protection? Deciding what's important to you is a question for every individual to answer on his own. The skeptical process can lead to the truth of what happened, but only you can answer what truths are important.

23. PAGANISM: A NAKED REBELLION

It's time to shed our arbitrary layers of corporate fabric and dance gaily through the forest glade wearing the suits we were born in — for the theme of this chapter is paganism.

Paganism is not well defined. The definition can be quite broad or progressively narrow. The broadest definition of paganism includes all religions but the Big Three: Christianity, Islam, and Judaism. To a member of the Big Three, a pagan can be anyone who is not a member of their particular church. As you tighten the definition, you first eliminate the Dharmic religions: Hinduism, Buddhism, Jainism, and Sikhism. Whittled down to just those who call themselves pagans, you have the Wiccans, Celtic Druids, witches, Goddess worshipers, and recreations of other ancient polytheistic religions like those from ancient Egypt, Greece, Rome, and the Vikings. For this reason, the term neopaganism is really more accurate to describe modern pagans. Neopaganism typically does not include any Satan worshipers, which is a bit of a popular misconception. In this chapter I'm going to use the term neopaganism as if it's a religion by itself, which isn't really correct, but should generally encompass the beliefs of most of those who consider their religion to be pagan.

Neopaganism is generally polytheistic, with gods ranging from divine beings to things in nature. Spiritualism and divinity are crucial aspects of neopaganism. Despite its separation from the world's major religions, neopagan faith is very much dependent upon supernatural beings or paranormal forces and energies. In some cases, neopagans have advertised their faith as a way to reject the inconsistencies and suspensions of science required by the major religions and yet still remain a spiritual person. However, this doesn't really hold water for me. The spiritual aspects of paganism are equally at odds with science. Pagan gods might be rocks or trees, or they might be Zeus and Athena, or they might be some other mystical force but they are still unmeasurable and undetectable paranormal entities. You can't have it both ways. If you maintain a belief in any spiritual entity, you are rejecting what science tells us about that entity.

Goddess worship is popular in neopaganism. The obvious question that the rest of us have is "Who is the goddess?" We've all seen the paintings of the dude with the beard, the white robe, and the Birkenstocks, but never of a goddess. The neopagan god and goddess are not necessarily specific beings. Many neopagans believe that whomever or whatever god is, is not necessarily knowable. But they also believe that the god has masculine and feminine aspects, which they call the god and goddess. Goddess worship is thus not the worship of a particular divine female being, it's a more general worship of femininity itself. Sometimes the goddess is linked to some of the ancient named gods like Athena, Ishtar, or Venus. Sometimes the goddess refers to divine spirituality that neopagans assign to maternity, fertility, and nurturing. Clearly the god and goddess concept is in direct contradiction with Christianity's Holy Trinity, so the absolute incompatibility of goddess worship and Christianity is an important distinction. This is another case where some neopagans try to have it both ways. But I'm not going to sit here and proclaim that this makes their religion invalid. Everyone is free to have whatever divine beliefs they want, and if they want to have a goddess that's compatible with Jesus or Mohammad, fine. It's no more or less valid than anyone else's concept of divinity.

One popular allure of paganism is its embracing of free sex and public nudity. I've always believed that more people secretly appreciate free sex and public nudity than are willing to admit it — at least when the chicks are hot. Wiccans have even institutionalized nudity, calling it "skyclad." This is good, because it better legitimizes the dress code for my hot tub. Skyclad apparel only, please.

Is there an obligation for those who are into skyclad self-expression and disestablishmentarianism to embrace the paranormal by joining a pagan religion? I don't see that there is. Go to Burning Man, if that's what floats your boat, or move to Los Angeles. You can have fun and indulge in individuality without adopting some form of supernaturalism. If the idea is to rebel against the straight lace church that your parents made you go to as a kid, rebel against it by recognizing that it's based on hooey rather than adopting some different but equally silly brand of hooey.

Another great way to buck the trend and be your own person is to use your own brain, by being rational and employing critical thinking, rather than using someone else's brain, and joining their organization, be it a neopagan religion, a radical environmental group, or a Republican campaign. Does the average modern Celtic Druid truly profoundly believe the doctrine of his religion, or does he just enjoy the company of a great group of people with a really neat philosophy? I'm all in favor of hanging out with great people with neat philosophy, even running around naked in the forest with them (especially if their chicks are hot), but I don't need to adopt belief in occult magic and reincarnation — fundamentals of druid doctrine — to do it. It would be great if joining them would give me magical powers, but rationality and critical thinking tell me that it would not be so. This has saved me many full moons of streaking through forests hoping for enlightenment.

Self expression, iconoclasm, impatience with social convention, and free thinking are all great things, and something that more people should engage in. But switching from one brand of hooey to another does not accomplish any of them, and doesn't indicate that your thought process was truly critical and skeptical, and certainly not independent or unique.

So while you're casting off your robe, cast off some of that joiner mentality and seek your own answers using your own brain.

24. Reflexology: Only Dangerous If You Use It

Let's lay back on the sofa, put our feet up, and receive a therapeutic foot massage, accompanied by the soothing sounds of the rainforest. Feel the energy as your body's impurities are cleansed, your wellness heals itself, and the cancerous tumor in your brain melts away — all because of this foot massage. We call it *reflexology*.

Reflexology is the art of rubbing the foot, with the belief that certain areas on the bottom of the foot are spiritually connected to parts of the body. Rubbing the part of the foot that correlates to the brain, for example, is supposed to cure anything that's wrong with your brain, like brain cancer. Rubbing the part of the foot that corresponds with your elbow is alleged to magically reconnect a torn elbow ligament. Developed in 1913 by a man named William Fitzgerald as "zone therapy", reflexology is based on the New Age definition of the word "energy". Fitzgerald believed that a mystical force field, not understood by science, that he called "bioelectric energy", ran through the body in ten vertical bands corresponding to your ten digits. Modern practitioners call Fitzgerald's mystical energy field "life force", and believe that adepts can manipulate this force field to promote any type of wellness in any part of the body, all through actions that correspond to a conventional foot massage.

Now, nobody disputes that foot massages do have benefits. They feel great, and absolutely promote relaxation and stress reduction. Unfortunately, these benefits can mislead people to conclude that the massage is working for whatever other malady is claimed to be treated. Another problem with reflexology is that, when used to diagnose a medical problem that does not in fact exist, the practitioner can claim that it is a future problem that's being diagnosed and treated. Time travel combined with medical treatment! If reflexology were to be tested and compared to the results of a real medical diagnosis, this time travel aspect allows its supporters to claim even a clean miss as a direct hit.

A listener to my podcast wrote in with the following letter:

> *Hello Mr. Dunning,*
> *I live in a small town in Iowa (pop. 4,000ish). About two years ago a fitness center was built (the Chickasaw Wellness Complex, CWC), which I think is pretty good for a town of our size. I've got a membership and have thus far been satisfied. However a (cover) story this week in our local paper was about the new 'Reflexologist' now employed at the CWC. I have attached the article.*
> *My issue is this: I would like to submit a letter to the editor refuting the article and exposing reflexology for what it is, pseudoscience. The reason I am emailing you then is that I need some help. I need some information and resources as well as talking points for my letter. Please help!*
> *p.s. Thanks for the podcast, I enjoy listening.*

And thanks to the listener for helping to fight the good fight and alerting the paper's readers to this sham. And here is the article that the *New Hampton Tribune,* in New Hampton, Iowa published:

> *What is Reflexology?*
> *As part of the Lighten up Iowa Kick-off Celebration held at the Chickasaw Wellness Complex on Thursday, January 4, Kabira Redcloud (not her real name), a Lay Minister of Reflexology, introduced area residents to Reflexology.*
> *According to information supplied by Kabira, Reflexology, or zone therapy, is the practice of stimulating points on the feet and hands, in the belief it will have a beneficial effect on some other parts of the body, or will improve a person's general health — helping a body heal from acute and chronic conditions, help reduce pain, stress and the effects of stress on the body such as high blood pressure.*

The most common form is foot reflexology. Practitioners believe the foot to be divided into a number of reflex zones corresponding to all parts of the body, and that applying pressure to tight areas of a person's foot will stimulate the corresponding body part, thus causing it to begin healing itself.

After a medical history assessment, in a Reflexology treatment Kabira first conducts a "Thumb Walk", pointing out tender areas of the bottom of a person's feet. These areas are documented and treatment is focused on them.

A machine called "The Drummer" is then used on the bottom and top of the feet, similar to massage machines. The Drummer can stimulate areas deeper and more effectively than fingers.

Kabira graduated from the Modern Institute of Reflexology with a 4.0 grade point average and recently became certified as a Lay Minister of Reflexology.

Now, at first glance, one reaction to this letter is that reflexology is probably pretty harmless, and this is the kind of New Age faith-based treatment that the majority of people seem to want these days. I've had foot massages on a number of occasions and they do feel pretty darn good, so I'm sure that the majority of Kabira's customers will come away feeling wonderful, at least until the massage wears off. Nothing wrong with that part of it at all.

But I wish that the *New Hampton Tribune* hadn't taken Kabira's press release so literally and reprinted it with so little reflection on its contents. What we have here is a newspaper advising its readers where to get a "medical history assessment" from a person with no medical training whatsoever. (If Kabira had any medical training, I'm sure she would have listed it on her resume before "Lay Minister of Reflexology.") This is absolutely unacceptable. From a liability standpoint alone, it's insanity for a newspaper to print this; and from an ethical standpoint it's egregious. The *New Hampton Tribune* has no excuse for stating that reflexology can improve a person's general health. Suppose a reader has a serious illness and goes to Kabira after reading this article, at the expense of time and money which could have been spent on crucial medical treatment. There is nothing in this article that suggests a patient should do anything else. And this is the central risk of reflexology: that a believer, or even a naive victim, will turn to reflexology in the belief that it can treat an illness, at the expense of proper medical treatment. This delay of treatment can result in serious injury or death.

I think my favorite part of Kabira's press release is that she trumpets her 4.0 grade point average from the Modern Institute of Reflexology. Notice that you'll find the Institute prominently listed on Wikipedia's "List of Unaccredited Institutions." Wow, a 4.0 GPA from an unaccredited correspondence school. The Institute has a web page describing the course of study to become a Lay Minister. The page consists largely of prayers, scriptural passages, and even a discussion of Biblical foot washing (for some reason, this particular institute mixes a large dose of Christianity in with Reflexology's usual paranormal claims). Sounds like a pretty rigorous medical course to me.

The article also states that Kabira is "certified" as a Lay Minister. Since any certification that she might have is from an unaccredited correspondence school about Biblical foot washing, and not from any medical board approved by the American Board of Medical Specialties, it's irresponsible of the newspaper to call her "certified" when she's offering what she calls a "medical history assessment". Kabira, and any other reflexologist who uses the word "medical", is about two inches away from prosecution for practicing medicine without a license, and any newspaper worth its salt should have refused to run her press release. In no way is any reflexologist certified to give any type of medical assessment, diagnosis, or treatment. To do so would be a felony.

New Hampton Tribune, clean up your act. People say "Don't shoot the messenger," I *am* shooting the messenger. Your article is irresponsible and endangers the health of your readers. Chickasaw Wellness Complex, what can I say. Offer massages, they're wonderful things. I haven't heard whatever you might be telling your customers, who are paying you for wellness, about reflexology — but I hope it's factual and contributes to their health, and doesn't put them at risk of seeking alternatives to needed medical treatment. And to the listener who wrote in, thank you for being the only voice of reason here and looking out for the health of your fellow Iowans.

25. SCIENTISTS ARE NOT CREATED EQUAL

You hear on the news all the time that scientists say this, scientists say that. For example, some friend of mine will try to convince me that the Earth is only 6000 years old because there are some scientists now supporting it. I often reply with something like "Sure, it's easy to find some whack-job who will say just about anything."

"No, no; not whack-jobs," they'll quickly say. *"Scientists."*

Oh. Well there are no whack-job scientists. News flash: Wherever you go, you'll find all kinds of people. All kinds of people, in every group. As if bearing the arbitrary, unsupported label of "scientist" means that you automatically know your ass from a hole in the ground. Does it?

What exactly is a "scientist", anyway? Is it someone with a degree in a scientific field? Is it someone who works in a scientific field? Is it someone who's won awards, or written articles in a scientific journal? Can it be a guy in his basement who has taught himself a great deal about a given subject? Can it be anyone who applies critical thinking to the world around him? Do you have to have the word "scientist" in your job title? Can someone simply call himself a scientist? Whatever it is, it seems that your word is cast in gold as absolute truth if someone refers to you as a scientist. Many people accept that too readily. If the 6:00 Action News team reports that a scientist says it, it must be true.

Not all scientists are people that we should listen to at all. Even the Nazi doctors who performed experiments on living humans during World War II were, by any practical definition, scientists. Would you want any of those guys telling you what's right and what's wrong? Nevertheless they held advanced degrees and were among Germany's top medical experts. It's weird to say it, and it's politically incorrect, but you can't disqualify Nazi doctors as valid scientists just because they were evil. Now go to the other end of the spectrum. Most people in the world — and thus, by extension, most people in the world with post graduate scientific degrees — attend religious services. The only thing that tells us is that those scientists do not apply skeptical critical thinking to the theological aspect of their lives. Beyond that, many of them are top experts in their scientific fields, Nobel laureates among them. You can't necessarily disqualify a scientist only because of certain aspects of what he does. Many detractors try to, but it's often not right. I'm considered a top expert in my professional field, and I absolutely have differences with most of my colleagues. Should I be cast out, or is it healthy to have diverse viewpoints within a community?

I submit that we shouldn't give any weight to someone's statements just because some person calls him a scientist. So then, what quality must a scientist have to be authoritative?

Maybe we should accept the word of a scientist if he has an advanced degree. Have you ever known an idiot with a degree? The fact is that practically any motivated person can eventually get any degree they want, if they're willing to put in the years. I'm sure that if James Randi wanted to, he could work hard and get a Ph.D. in Divinity from Oral Roberts University. The reverse is also true: A staunch Creationist could no doubt become a doctor of astrophysics — indeed, many astrophysicists out there undoubtedly are Creationists. Thus, when you hear a Creationist defend his position by quoting from a scientist (name any astrophysicist) who believes in it, that hardly means that the entire science of astrophysics has concluded that the universe was created by a magician. Not only is the fact that someone holds a particular degree not a reliable indicator that he is an expert in that field, many degrees are themselves pretty worthless as indicators that the holder has a scientific mind. Legitimate accredited Ph.D.s are available in many fields not associated with science, such as divinity, philosophy, dance, or fiction. Many people can go around rightly calling themselves a doctor, but having no scientific background at all. Really the only thing a degree tells you about someone is where they drank themselves into a stupor when they were 19. I refer you to my own Ph.D. on ThunderwoodCollege.com. Is a scientist automatically qualified because he has an advanced degree? No.

Maybe we should accept the word of a scientist if he works in a certain industry. Have you ever had a boss who didn't know as much as you? Have you ever worked with someone who hated his job or didn't care about it? Think about the company where you work right now, and think of that one guy in the office that everyone thinks is a kook. Is he a kook for a reason? There may be people at your company who would make good representatives of your work if you put them in front of a group to speak. Are there also people at your company that no way would you want them representing what you do? Is a scientist automatically qualified because he works in a certain industry? No.

The fact is that calling someone a scientist doesn't mean that he's smart, that he's right, that he thinks scientifically, or that he's anything more than a waste of space. You can't easily qualify someone just because they're called a scientist, and you can't easily disqualify a scientist because of some stuff that he does. All of this means that the label of "scientist" is pretty darn worthless by itself. When you hear any claim validated by the fact that some "scientists" support it, be skeptical. You need to know who they are, what their interest is, and especially what the preponderance of opinion in the scientific community is. You need to know if the scientist being quoted actually has anything to do with this particular subject, or if his specialty is in an unrelated field. Look to see if this scientist has authored a good number of publications on the subject in legitimate peer-reviewed journals. Find out what other published scientists in his field say about him. Determine whether his views are generally in line with the preponderance of opinion among his peers in his discipline. Fringe opinions are on the fringe for a reason: they're usually wrong.

26. The Magic of Biodynamics

Join me now as we pull a cork and pour ourselves a glass of the latest fashion in winemaking: biodynamic wine.

If you're not familiar with biodynamics, the best way to think of it would be as a magic spell cast over an entire farm. Biodynamics sees an entire farm as a single organism, with something that they call a life force. Believers say that this life force can be increased, thus improving crop quality and health, by following conventional organic methods plus the application of a special magical potion. The potion is not intended to have any direct physical effect — its intent is to charge up the life force of the farm. The word *biodynamic* literally means life force, according to its Greek roots *bio* and *dyn*.

Rudolf Steiner was a philosopher, a multi-faceted artist, a playwright, and a self-described clairvoyant. He gave a series of eight lectures in Germany in 1924, which became the essential bible of biodynamics. To Steiner's credit, he always insisted that his students test everything he said and not take it at face value; but to his detriment, he did no testing of his own, but rather described the methods to be followed based only on his own inspiration. Steiner's eight lectures are available on the Internet, and if you're curious you should look them up and read them. They start with some conventional sounding discussion of soil chemistry and nutrients, but then devolve into a progressively more vague lecture about non-physical beings and elemental forces.

Now you probably think I'm coming across as overly cynical or that I'm deliberately painting biodynamics in a negative or irrational light. I don't want you to feel that I'm coloring this in any way, or at least in any undeserved way, so please draw your own conclusions from what I'm about to read. This comes directly from a prominent biodynamics web site, so this is their description, not mine. The potion consists of nine ingredients (or preparations, as Steiner described them), numbered 500 through 508. Here are their descriptions and the instructions:

> *500: A humus mixture prepared by stuffing cow manure into the horn of a cow and buried into the ground, 40-60 cm below the surface, in the autumn and left to decompose during the winter.*
>
> *501: Crushed powdered quartz prepared by stuffing it into a horn of a cow and buried into the ground in spring and taken out in autumn. It can be mixed with 500 but usually prepared on its own (mixture of 1 tablespoon of quartz powder to 250 litres of water) The mixture is sprayed under very low pressure over the crop during the wet season. It should be sprayed on an overcast day.*
>
> *Both 500 and 501 are used on fields by stirring the contents of a horn in 40-60 litres of water for an hour and whirling it in different directions every second minute.*
>
> *502: Yarrow blossoms stuffed into urinary bladders from Red Deer, placed in the sun during summer, buried in earth during winter and retrieved in the spring.*
>
> *503: Chamomile blossoms stuffed into small intestines from cattle buried in humus-rich earth in the autumn and retrieved in the spring.*
>
> *504: Stinging nettle plants in full bloom stuffed together underground surrounded on all sides by peat for a year.*
>
> *505: Oak bark chopped in small pieces, placed inside the skull of a domesticated animal, surrounded by peat and buried in earth in a place where lots of rain water runs by.*
>
> *506: Dandelion flowers stuffed into the peritoneum of cattle and buried in earth during winter and retrieved in the spring.*
>
> *507: Valerian flowers extracted into water.*
>
> *508: Horsetail.*
>
> *One to three grams (a teaspoon) of each preparation is added to a dung heap by digging 50 cm deep holes with a distance of 2 meters from each other, except for the 507 preparation, which is stirred into 5 litres of water and sprayed over the entire compost surface. All preparations are*

thus used in homeopathic quantities, and the only intent is to strengthen the life forces of the farm.

It doesn't say so but I think you're supposed to chant "Double double toil and trouble, fire burn and cauldron bubble" while preparing it.

Notice that the potion is applied to the compost in the ratio of one sixteenth of an ounce per *ten tons* of compost. Biodynamicists describe this as a homeopathic amount, which means diluted to virtually undetectable levels, which certainly applies in this case. But again, the potion is not intended to have any direct physical effect. It's there to stimulate the life force. I was not able to find any proposals describing what the mechanism for this effect is alleged to be, or exactly what the effect is, other than the vague phrase "strengthen the life force."

If you're wondering why cows' horns figure so prominently in the potion, this explanation comes from one of Steiner's lectures:

> *The cow has horns in order to reflect inwards the astral and etheric formative forces, which then penetrate right into the metabolic system so that increased activity in the digestive organism arises by reason of this radiation from horns and hoofs.*

Remember, Steiner encourages you to test this scientifically. All you'll need is an astral and etheric formative force gauge. I think they have them at Radio Shack.

Frankly, I'm shocked to see adults in industrialized nations mixing magic potions in the twenty first century. If they were doing it for ceremonial reasons, or to honor some folk tradition, fine; but doing it earnestly in a sincere effort to improve a harvest is alarming. Essentially, we're talking about witchcraft and sorcery being employed as a modern farming tool, today, in the United States. Go to your local liquor store and you'll find wines from biodynamic farms. Pick up a wine magazine and you'll find reviews of biodynamic wines. This phenomenon has already infiltrated our daily lives. Is this a commentary on the state of our educational system? Does it reflect a change in what parents are teaching children?

I'm not persuaded by the anecdotal evidence either. What you'll usually hear from biodynamicists is something like "I know it sounds weird, but the fact is the wines are actually better, and that's an undeniable difference that you can taste." I'm sure there actually *is* a difference that you can taste. No two wines *ever* taste alike. Grapes from every vineyard in the world taste different, and grapes from the same vineyard taste different each successive year. That's why there are good vintages and bad vintages. Any experienced vintner making the same wine from the same grapes in the same vineyard will easily be able to tell you which vintage a given bottle is from. Whether the growing method changed from conventional to organic to biodynamic or to anything else, the next batch of wine *must* taste different. That's a fact of winemaking. If ever there was a science where a valid controlled trial was absolutely impossible, it's winemaking. Anecdotal reports that wine quality improved after the farm became biodynamic can't be given any credence. Not only is "improved" a completely subjective matter of opinion that will differ among all wine drinkers, but any chromatographic analysis of the chemical content of the wine must be different year after year, whether the vineyard cast a magic spell over their compost or not. Neither any actual effect nor any causal relationship can be evidenced.

No doubt there are valid marketing reasons for making and selling biodynamic wines or other crops — there are always going to be customers who want it. And that's fine. But please don't teach your children that magic potions and non-physical beings are the way to achieve success in agriculture or anything else.

27. Chemtrails: Death from the Heavens!

It's time to put on our Men In Black suits, buy a plane ticket, and spray the world with mysterious chemicals as part of an evil government conspiracy. For the subject of this chapter is chemtrails.

Wow. Where to begin. I read a fair amount of skeptical, paranormal, and conspiracy web sites, but I don't recall ever reading so much vituperation, anger, and name calling as when I read a few forums discussing chemtrails. If you're not familiar with the term, chemtrails are what some conspiracy theorists call aircraft condensation trails. Most of them don't believe that conventional contrails exist, and that when you see one, you're actually seeing a trail of mysterious airborne chemicals sprayed from the aircraft. Those who do concede the existence of contrails often claim subtle differences in appearance or behavior between a condensation trail and a chemical trail.

First, let's discuss exactly what a contrail is. A condensation trail, also called a vapor trail, forms at altitudes above 25,000 feet in temperatures below -40 degrees when engine exhaust condenses into ice crystals, creating an artificial cirrus cloud. Water is produced by hydrocarbon burning engines in about the same quantity as fuel consumed, and in the right conditions, this extra addition of water into the air pushes the water vapor past the saturation point, and condensation occurs. It takes a moment to happen, which accounts for the contrail appearing a short distance behind the aircraft, rather than immediately, like you'd see from a smoke pod on an aerobatic plane, or like when a crop duster releases chemicals. Contrails can also be caused at high altitudes by the extreme low pressure areas created by wingtip vortices, which reduce the temperature enough to condense the existing moisture in the air. As previously mentioned, many chemtrail believers claim that there is no such thing as a condensation trail. Since they're well understood, 100% reproducible, and observable practically any time you look into the sky, the onus is really on the believers to prove that no such thing is possible. In my opinion they have quite an uphill battle on this one.

By the way, in case you're wondering whether I meant Celsius or Fahrenheit when I said -40 degrees, here's a bit of science trivia for you. -40 is the point where the Celsius and Fahrenheit scales are the same. Can you figure out where Fahrenheit and Kelvins intersect?

Some chemtrail believers say that the population is being gassed with some unknown chemical for an unknown reason. Others tie into pop culture, suggesting that chemtrails are the active manifestation of one proposal to combat global warming by placing dust into the upper atmosphere to reflect sunlight. One guy told me "You guys are so full of yourselves just can't conceive of spiritual warfare can you? Try this fact: malevolent interdimensional entities are involved — project keep the human slave race suppressed."

Like all conspiracy theories, chemtrails require us to accept the existence of a coverup of mammoth proportions. In this case, virtually every aircraft maintenance worker at every airport in the world needs to be either part of the conspiracy, or living under a threat from Men in Black, with not a single whistle blower or deathbed confession in decades. Yet the threats are not frightening enough to prevent stories like the following from appearing on the Internet:

> For reasons you will understand as you read this I can not divulge my identity. I am an aircraft mechanic for a major airline. I work at one of our maintenance bases located at a large airport. One day last month I was called out from our base to work on a plane for another airline. When I got to the plane I found out that the problem was in waste disposal system. When I got into the bay I realized that something was not right. There were more tanks, pumps, and pipes then should have been there. As I tried to find the problem I quickly realized the extra piping and tanks were not connected to the waste disposal system, at all. To my

amazement the manuals did not show any of the extra equipment I had seen with my own eyes. I even tied in to the manufacturer files and still found nothing. I could trace the control wires from the box to the pumps and valves but there were no control circuits coming into the unit. The only wires coming into the unit was a power connection to the aircraft's main power bus. The system had 1 large tank and 2 smaller tanks. It was hard to tell in the cramped compartment, but it looked like the large tank could hold about 50 gallons. The tanks were connected to a fill and drain valve that passed through the fuselage just behind the drain valve for the waste system. I discovered that the pipes from this mystery system lead to every 1 out of 3 of the static discharge wicks. These wicks had been "hollowed out" to allow whatever flows through these pipes to be discharged through the fake wicks.

About 30 minutes later I was paged to see the General Manager. When I went in his office I found that our union rep and two others who I did not know were waiting on me. He told me that a serious problem had been discovered. He said that I was being written up and suspended for turning in false paperwork. I sat at home the first day of my suspension wondering what the hell had happened to me. That evening I received a phone call. The voice told me "Now you know what happens to mechanics who poke around in things they shouldn't. The next time you start working on systems that are no concern of yours you will lose your job! As it is, I'm feeling generous, I believe that you'll be able to go back to work soon." CLICK. Well that's it. Now I know 'THEY' are watching me.

Another forum poster replied:

I, too, work for an airline, though I work in upper management levels. I wish I could document everything I am about to relate to you, but to do so is next to impossible and would result in possible physical harm to me. Airline companies in America have been participating in something called Project Cloverleaf for a few years now. The earliest date anyone remembers being briefed on it is 1998. I was briefed on it in 1999. The few airline employees who were briefed on Project Cloverleaf were all made to undergo background checks, and before we were briefed on it we were made to sign non-disclosure agreements, which basically state that if we tell anyone what we know we could be imprisoned. About twenty employees in our office were briefed along with me by two officials from some government agency. They didn't tell us which one. They told us that the government was going to pay our airline, along with others, to release special chemicals from commercial aircraft. When asked what the chemicals were and why we were going to spray them, they told us that information was given on a need-to-know basis and we weren't cleared for it. He seemed perturbed at this question and told us in a tone of authority that the public doesn't need to know what's going on, but that this program is in their best interests. He also stated that we should not tell anyone, nor ask any more questions about it.

I take back my comment about the lack of whistle blowers. It seems that just about everyone is discussing it freely, and since the government is openly laying out virtually all the essential details to 20 people in every airline office in the world with complicity from all the union reps, why the secrecy? If we're truly on the edge of a global warming catastrophe and a worldwide cooperative effort to spray reflective dust into the atmosphere is the only way we can save ourselves, wouldn't a little *glasnost* be a more effective approach? Of course it would, but then we wouldn't have had those two very exciting and original stories.

If the goal was to drug a population, you'd put the gas where it could reach people, you wouldn't release a minute quantity into the jet stream where it's going to be diluted to undetectable levels and eventually rain into the oceans. If the goal is to put reflective dust into the atmosphere to fight global warming, you'd need to plant it a lot higher than commercial jets can travel if you don't want it to rain right back to earth (in fact you need to put it well up into the stratosphere, not just at the top of the troposphere), and you'd need a lot more than can ever be delivered with 50 gallon tanks. But then, the arguments in favor of chemtrails don't appear to be based on any kind of logic. They appear to be based on nothing more than imagined conspiracies and sightings of visible contrails, which are well understood, and are being left by planes every hour of the day over every city in the world, and have been for over 50 years. It's how the malevolent interdimensional entities remind us of their ominous presence.

28. NATURAL HYGIENE: HEALTH WITHOUT MEDICINE (OR WISDOM)

In this chapter we're going to flush our entire medicine cabinet down the toilet and try "natural hygiene," the practice of improving our health by avoiding medical care.

The hypotheses behind natural hygiene suggest that modern medicine and vaccinations are harmful to the body, and that viruses, bacteria, and germs are not harmful. Basically, take everything that modern science has taught us about the human body, turn it upside down and backwards, and there is your natural hygiene. However they do recognize one fact of biology, and that's that the human body has the power to heal itself. But they don't really understand what this means: They believe that the *only* way a human body can be healed is on its own, without medical care. For example, if you have an infected wound, natural hygiene suggests that a shot of penicillin will actually make things worse. In fact, such wound care as this can sometimes be the only thing that will save your life. Many practitioners do bend their own rules in cases of trauma or emergency care, acknowledging that medical care is actually helpful in an emergency. It's the rest of time, normal wellness or treatment of chronic illness or disease, where they believe medical care is counterproductive.

The human body does have amazing recuperative powers. Its immune system is powerful and sophisticated. Every day, someone's immune system manages to overcome some disease that's usually fatal. And it's these relatively few lucky victories that always get all the attention. When a person who doesn't understand medicine reads in the National Enquirer that someone overcame cancer while trying natural hygiene, it's natural to assume a causal relationship. In fact, as we know from medical history, more people who treat their cancer will survive than those who don't. It's the exceptions that make the headlines, and comprise the bulk of the anecdotal evidence supporting natural hygiene.

When a practitioner of natural hygiene cuts his finger and sees it heal, he attributes this to his natural hygiene lifestyle. Really this is just the body's normal process. If he'd put on some Neosporin and a band-aid, it probably would have healed quicker and with less risk of infection.

Prior to 100 years ago, you were actually better off not going to the doctor if you became ill. The doctor was likely to bleed you, or induce vomiting, in order to balance your humors. Illness was thought to be caused by an imbalance in the four basic humors: blood, phlegm, black bile, and yellow bile. Needless to say, this level of treatment didn't get you very far. You were just as likely to die from infection caused by the bloodletting incisions.

Since then, with the advent of modern medicine and a century of its development, we've doubled the average life expectancy in the Western world from under 40 to almost 80. Most of this gain in average life expectancy has come from reductions of infant mortality and early childhood illnesses. Generally, if you can survive early childhood, you have a good chance of reaching middle age or even older.

Modern-day non-civilized native tribes, who lack access to modern medicine and are the only groups currently practicing natural hygiene in large numbers, have an average life expectancy of just 34 years at birth. But this doesn't mean that everyone drops dead at 35, like some jungle version of Logan's Run. Those lucky enough to survive into their teen years have an average life expectancy of almost 60. What this means is that natural hygiene practitioners are at greatest risk of death during infancy and early childhood. Without inoculations and infant care, many children die and bring down the whole average.

Modern Western practitioners of natural hygiene are people who make the choice sometime during healthy adulthood. This means that they have generally already received their inoculations at an early age, and it obviously means that they already survived infancy and early childhood. So, doing nothing else, and having already been brought into adulthood through modern medical care, their human genes already provide them quite a long lifespan. In fact, since infant mortality brings down the whole average for everyone, an adult's life expectancy is already higher than the average life expectancy. This simple mathematical curiosity accounts for the fact that natural hygiene practitioners can generally claim to live longer than average. However, if they were required to place their bets just before childbirth instead of 20 years later, their average would be no higher than the general population, and probably less since they will not accept treatment for later stage illnesses like heart disease and cancer.

Natural hygiene practitioners do follow some very good health practices. They generally don't smoke or drink alcohol excessively, and they often follow a healthy low-calorie diet. These are all great practices, but they can be followed by anyone. You don't need to add the weird element of shunning healthcare to enjoy the benefits of these simple healthy choices.

29. Orbs: The Ghost in the Camera

Next time you pick up a camera, watch out. You're holding in your hand the very device responsible for tens of thousands of the most bizarre and unexplainable type of ghost photographs: Orbs.

Orbs, formally called Spirit Orbs, are those semitransparent white balls seen floating around in many photographs taken in ghostly locations. Orbs are among the class of paranormal phenomena that are visible only to cameras, and not to the naked eye.

The usual hypothesis presented by believers is that orbs represent spirits of dead people, though some support variations on that. The science behind this hypothesis is not clear. For example, there are no plausible hypotheses that describe the mechanism by which a person who dies will become a hovering ball of light that appears on film but is invisible to the eye. There are lots of other things that a dead person might become, presumably; and the only reason believers have chosen orbs seems to be that orbs are the most common unexpected objects seen in photographs. If there was any good science behind this, there would at least be some plausible proposals for what the orb might consist of, how and why it is generated by a dead body, why it floats in the air; and also some good predictions about who will become an orb after they die, what size and color that orb would be, and where and when it can be found. I welcome any hypotheses that would explain how orbs could be a real phenomenon, but I haven't been able to find any. The only evidence is anecdotal reports and, of course, the obligatory photographs, found on the Internet by the thousand.

Orbs most often appear on camera when a piece of airborne dust, an insect, or a water droplet is close to the camera, outside of the depth of field, and the flash source is no more than a few degrees away from the axis of the camera lens. This causes the object to be brightly light but way out of focus, resulting in a semi-transparent whitish circle. If the flash or other light source is significantly off of the axis of the lens, you won't get nearly as much light reflected right straight back to the camera. If the object is within the depth of field it will be in focus and generally very small, and probably not noticeable. If the object is not very close to the camera, again it won't pick up enough light from the flash.

I'm often challenged by believers that if orb photos are so easy to take, why don't I do it then? I don't because many people have already done so. If you want great step-by-step instructions for taking an orb photo, go to assap.org and click on Paranormal Photos. You will get all the examples, instructions, and explanations that you could ask for. I do have a couple of orb photos that I took by accident inside an abandoned mine shaft — doubtless the hapless spirit of a murdered miner — and you can find those pictures online at Skeptoid.com.

Now, it would not be correct to state that orb believers don't accept this explanation. Most actually do; in fact, many web sites that archive ghost photographs no longer accept orb photographs, with the explanation that orb photos too often show false orbs produced by the photographic effects described above. Nevertheless, most believers still feel that there are legitimate orb photos that do show ghosts or spirits or energy or whatever they want to call it. One differentiator that I've heard several times is that a false orb photo will have a blue edge, while a real orb photo, showing spiritual energy, will not. Once again, there's a simple explanation that's well known to photographers. Basically, cheap optics and certain sensors will produce this blue edging. To see some examples, go to a high end digital camera review web site, such as dpreview.com, and look through some of the photographic tests of cameras that they review. A good place to see the variance of cameras producing blue edges is the resolution test chart. The effect can also be caused artificially by the camera's image processing software when certain luminance and chrominance settings are in effect. Finally it can also occur with even a high-end camera with the right white balance adjustment when using a flash of a certain color temperature. In short, blue edges on orbs can be added or subtracted by the camera, and often are, and so should not be considered a reliable indicator of whether a given orb is actually a ghost.

I've also found statements attempting to debunk the evidence that orbs are caused by the flash reflecting from dust outside the depth of field. These claims are based on multiple successive photographs taken immediately one after the other, where an orb appears in one but not in the others. Presumably, if there's dust in the air, there's dust in the air; and it's not going to float away in a split second. Fair enough. But I've never seen such a series of photographs. Thus this is purely an unsubstantiated anecdotal report, from someone who probably has an agenda — judging by their pro-orb web site. I find it hard to believe that a dust particle would remain in exactly the same place for the second or so that a fast digital camera would require to take two pictures. It only needs to move half an inch or so if it's close to the lens, and even the movement of your hand on the shutter will make enough wind to move it. The slightest breeze or air current would move it well out of the way. And even in a perfectly still room, Brownian motion is by itself more than enough to make that dust ancient history by the time the camera takes a second shot.

Another hypothesis about orbs is that they are not the spirit at all, but rather energy being transferred to a spirit. Suppose that a spirit is hanging out near a power source, be that a person, a powerline, a warm fireplace, or something else. The spirit, by its nature, draws energy which moves into it in the form of glowing, hovering balls. I read from one source that supports this idea that "the laws of physics say that energy transferred like this would naturally assume the shape of a sphere." Hmmmm. Refer back to the chapter on New Age Energy. Energy is not a hovering, glowing, physical substance that goes around and does things. Energy is simply a measurement, so it's hard to imagine what law of physics he was talking about. Perhaps these hypothesizers mean to say that the spirit is drawing heat, or electricity. Well, neither heat nor electricity are ever seen to move around in the form of gently hovering transparent white balls. And don't say "ball lightning", because you bet your ass that every ghost hunter in the room would know if there was a wicked ten million volt ball of death banging around; they wouldn't have to wait until they got home to check their film to find out about it.

So, in conclusion, I basically came up short seeking a plausible hypothesis for the existence of orbs. Until someone comes up with one, I'm satisfied that the evidence shows orbs to be merely a well understood and commonplace artifact of photography.

30. Raw Food: Raw Deal?

Turn your stove off, put those pots and pans away, and prepare to be dazzled by a delicious meal made from fresh, raw fruits and vegetables. But wait! They're not *just* healthy and delicious: Raw foodists are claiming a lot more than that.

Raw foods are delicious, they're easy to prepare, you'll probably spend less money on food, you'll spend less time washing dishes, and you're a lot less likely to burn yourself in the kitchen. Nobody denies any of these benefits; I certainly don't. What I do deny are some of the other, less honest claims that some of the more militant raw foodists make. I don't know why they make these claims, since most of them are so obviously untrue. Raw food is already a great thing; it doesn't need to be defended or supported by lies.

I should point out that I am a *huge* raw food fan. Nobody likes raw food more than I, or eats it in larger quantities. However, my raw food diet consists mostly of fish in Japanese restaurants, and it only supplements my regular diet of normal food. But by making this disclosure, I do claim full protection from the appearance of bias.

Let us now listen to some of the disingenuous claims made by some raw foodists:

Animals in the wild don't get sick because they eat a natural, raw diet.

Apparently, a lot of raw foodists honestly believe — or at least say that they believe — that wild animals don't get sick, and therefore we wouldn't either if we only ate raw food like animals, as if *that's* the significant difference between people and wild animals. In fact, disease is the major cause of mortality in wild animals. Check with any of the various wildlife groups who send veterinarians and biologists out into the jungles of the world to care for wild animals. This claim is especially hard to support given the numerous high profile cases of disease in wild animal populations: avian flu, chimpanzee Ebola, widespread hoof and mouth disease, bubonic plague in rats. The corollary claim is that raw foodists will not get sick. Unfortunately there is not a single example in the whole human race of someone who doesn't get sick, so this claim is just as untrue. Of all the original raw food gurus who have since passed away, please note that they have all passed away.

Eating raw foods will increase your lifespan.

Humans are living longer than ever before, and that's shown through overwhelming evidence. This is due to modern health care. You can try to debate this point and say that cooked food makes us live shorter than we naturally should, but you're on thin ice. All the evidence is completely against you. The argument is purely faith-based.

By far the greatest driver in longevity is heredity. Diet is not a significant factor, statistically. Actually there is some recent research showing that 80% of centenarians have abnormally large HDL particles, compared to 8% of the general population.

On a related note, I am intrigued by the low-calorie longevity. As you probably know, lab mice live 50% longer when fed an extremely low-calorie diet. There is a group of people who call themselves "life extensionists" who eat low-calorie diets. It should be noted that the significant factor in this diet is low-calorie: not vegan, not raw, not organic, not free of corporate hate energy; simple low-calorie is all it takes. But since there are not yet any long term clinical studies on humans, we can't yet assess whether what's good for mice is good for people, but it is still interesting.

Because of the same mathematical curiosity that we discussed in the Natural Hygiene chapter, people who adopt a raw food diet can genuinely claim to have average lifespans that are longer than the general population. This is because most raw foodists choose to adopt the lifestyle during healthy adulthood, when they're already past infancy and early childhood where many deaths occur in the general population, thus bringing down the general population's average life expectancy. When any group composed largely of adults claims a longer life expectancy than the general population, be skeptical of the reasons they give. It's true of all adults.

Humans are the only primates who eat meat.

You'll hear this a lot from raw foodists, but it's simply not true, as a read of any reference material, or a trip to any zoo, will reveal. Almost all apes are omnivorous. At one extreme you have baboons, who have been known to hunt goats and sheep in packs. At the other extreme you have gorillas, who eat insects as a small part of their diet. Most other apes, such as chimps, eat eggs, birds, and small mammals. Anyone who tells you this is trying to support an unsupportable claim: they're simply telling a far-out lie to convince you that eating meat is unnatural.

Cooked food is toxic.

I'm not even sure how to answer this one. Obviously, if cooked food was toxic, everyone on earth would have died long ago. Generations ago. Tens of thousands of years ago. Every speck of evidence shows quite conclusively that everyone talking about this is, well, alive. Cooked food is not toxic, or else we'd be dead.

Cooking makes organic compounds non-organic.

Let's review what an organic compound is. Ever take o-chem in college? Organic chemistry is the study of carbon compounds, and organic compounds are those formed by living organisms, with molecules containing two or more carbon atoms, linked by carbon-carbon bonds. These can be double bonds, where the carbon atoms share 4 electrons, or in the case of saturated fatty acids, they can be single bonds, where the carbon atoms share two electrons, and the other electrons are shared with bonded hydrogen atoms. Breaking these bonds would, in effect, make an organic compound non-organic.

So really, the claim being made by the raw food people is that cooking breaks those carbon-carbon bonds. You would have to really, *really* cook your food to break these bonds. Carbon-carbon bonds will begin to break at temperatures above 750 Fahrenheit, or about 400 Celsius. So if you cook your food in a ceramics kiln, then yes, it is possible to chemically change it into a non-organic compound. But if you're looking for it to happen at regular cooking temperatures, well then, you need to retake your o-chem.

Cooking kills needed enzymes in the food. Without these enzymes, the body cannot properly digest the food.

This is one of the more common claims that you'll hear, and it's based on a gross misunderstanding of digestion basics. Our digestive enzymes are produced by our body, and secreted into our mouth and stomach through glands. Humans do not need to eat digestive enzymes in order to digest. We are not Jeff Goldblum in *The Fly.*

Moreover, the claim that cooking "kills needed enzymes" is silly on two fronts. First, pretty much anything that you digest gets "killed" in the process — that's kind of the whole point of digestion. Enzymes contained in the food that you eat are broken down into their constituent amino acids by your digestive process, and it's these amino acids that are absorbed through your intestines. "Killing needed enzymes" is pretty much what you want your digestive system to do: break them up into amino acids that you can use. Trumpeting this fact with alarmism as if it's a bad thing, should really give you pause to consider how well these people know what they're talking about.

White blood cells flood the stomach after eating cooked food, because they're trying to fight the poison that just entered the body.

Neither I nor a doctor friend had ever heard of a mechanism by which white blood cells could enter the stomach, except in certain acute conditions involving gastric ulcers, where the stomach is open to the vascular system. The claimed phenomenon of pus in the stomach as a consequence of consuming cooked food does not ever appear to have been observed in medical literature, so this has all the appearances of being just another made-up lie.

Cooking food renders it unrecognizable to the body as food.

Well, again, we have a misunderstanding of the digestive system. The word "food" is just a label that people put on certain things that they eat. I could also eat dirt, and I wouldn't call it food. My stomach doesn't care what I call it: The digestive enzymes in my saliva and my stomach are going to treat it just the same. A raw food guy might eat a banana and a steak, but he's only going to label one of them food. I label both of them food. Doesn't matter to the digestive enzymes, which make no such query and apply no such labels. Regardless of what you call it, the digestive enzymes are going to break down whatever parts of it they can, and the resulting molecules are still going to be absorbed through the epithelial cells in the ileum. The dirt's going to pass right through me, but that cooked steak is going to end up in the raw food guy's bloodstream, nourishing his body, whether he likes it or not. He doesn't have to call it food.

Animals live to a much greater multiplier of their maturity age.

For some reason, this is often put forth as support for raw foodism. Presumably, raw foodists blame human consumption of cooked food for this. Why they draw this particular causal relationship is not clear.

And again, it's another claim that's simply not true, and you don't have to dig very far to discover this. Say that humans reach sexual maturity around age 13, and our average lifespan is around 75. That's about a 6:1 ratio of lifespan to maturity age. By the raw foodists' claim, animals should all be "much higher" than this. While it is true that most smaller animals are higher, most larger animals are right in the ballpark with us, and in many cases their ratios are lower. A lot of animal species achieve sexual maturity when they reach a particular size, not a particular age, so you can't always draw a direct comparison. Bottlenose dolphins, for example, range from 2:1 to 4:1, meaning that they live to be only two to four times their maturity age. Elephants are the same as humans, about 6:1. Gorillas are a little less than humans, ranging 5:1 to 6:1. Siberian tigers range from 4:1 to 5:1. Grizzly bears range from 3:1 to 6:1. In short, the claim is patently untrue, in addition to being irrelevant to raw foodism. The ratio of lifespan to maturity age has evolved in each species by the normal process of natural selection, adapted for each species' needs and environment. Evolution is not driven by recent diet choices.

If you want to eat raw food, by all means, go right ahead. It's healthy and it's delicious. But you can enjoy it without making absurd claims, and you can enjoy it without pointlessly attacking the alternatives. Please be careful that you don't catch *E. coli* or salmonella, unless you're one of the people that tries to spread any of the above lies; in which case, don't worry about *E. coli* and salmonella. If it's raw it can't hurt you. Eat up.

31. THE DEVIL WALKED IN DEVON

Snow fell all through the wee hours of February 8, 1855, in Southern Devon, a county in south western England.

When the sun rose at last, villagers throughout the county awoke to find a set of strange footprints stretching over 100 miles throughout the county. But it wasn't the length of the track or its sudden appearance that caused the most alarm. It was that the track at one point went right through a 14-foot-high wall, leaving untouched fresh snow on top of the wall. Elsewhere, the track went through a haystack, emerging on the other side. It was also found to enter a 4-inch drainpipe, and continue out the other end. In some places the track stopped inexplicably, only to reappear elsewhere. Most significantly, the track was found to have crossed a two mile stretch of river, picking right up again on the other side as if the maker of the footprints had walked on water. Theories and explanations abounded, until some clergymen suggested that perhaps the devil was on the prowl. From that moment on, the devil was said to have walked in Devon.

When people hear this story they generally imagine horseshoe-sized cloven hoof prints. However, this was not the case. All reports are that the footprints were between one and a half and two inches in size, and only eight inches apart. If this was the devil, he must have been more petite than he's usually depicted, and up on ballet pointe the whole way. No photographs exist but there are some detailed drawings of what the footprints looked like: basically, horseshoes. Now you've seen snow before, and you can imagine what a one and a half inch long footprint would look like in several inches of new fallen snow. Would it really be detailed enough to get a clear horseshoe drawing? Of course not. One suggested explanation was a field mouse, which does hop (often when there is snow that it can't easily walk through), and which does leave footprints in a V pattern. Each hind foot is long and narrow, and when they hop, their heels are together. A V is not so dissimilar from a horseshoe that the possibility cannot be discounted, given how indistinct the footprints would be in soft new snow. Some accounts describe it as a heavy snowfall, and others say it was a light snowfall. Now, it does get quite cold in Devon county and they definitely do get snow, but it's seldom more than a few inches and it certainly doesn't last all winter. This episode was in February, so a heavy snowfall is reasonable, though heavy is a relative term. Several inches is probably a good estimate. A single snowstorm is not likely to cover more than 100 miles, but we don't really have enough information about the preceding days to know whether there might have already been enough coverage over the area in question.

Among the possible culprits nominated by the witnesses were some kangaroos known to have escaped from a private owner, a Mr. Fische. Although the track sounds too small for kangaroos, the size and age of Mr. Fische's kangaroos was not published, so possibly they were quite young, or they may have even been a smaller species, like one of the smaller species of wallaby, some of which are no larger than small rabbits when full grown. This explanation does have one strong piece of evidence in its favor: the fact that kangaroo tracks were almost certainly unfamiliar to residents of England in 1855. Given that this was a rural area, and snowfall was frequent, tracks of commonplace animals would have been easily recognized. Whether Mr. Fische recovered his kangaroos is not known.

The element of this story that does it the least credit is the claim that this track stretched for over a hundred miles. In 1855 the means didn't really exist in Devon to travel a hundred miles in a single day to verify the length of this track, especially when the way is obstructed by two mile stretches of water. Devon is quite rural: farmlands, brooks, forests, hills; and must be teeming with small wildlife. Small rodents would have been running around everywhere, leaving tracks that would be pretty hard to tell apart from the description of the devil's footprints. For a group of investigators to follow one set of tracks to the river, leave their task long enough to find a bridge or a boat, cross, resume the search and spend enough time as it would take to find a set of mouse-sized footprints two miles away from where they were last seen, and all of this while plenty of other mouse-sized forest creatures are hopping around through the snow leaving tracks, and yet be certain that they found the one and only set of devil footprints, strains credibility.

A prominent biologist named Richard Owen declared the tracks to be those of a badger, perhaps distorted by freeze-thaw action. Other nominations included raccoons, rats, swans, and otters. And, of course, the most famous nomination: that of the devil himself.

What of these feats of moving through walls and haystacks, walking on water, and moving through drainpipes? I find it hard to give these much weight. These are purely anecdotal, unsubstantiated, uncorroborated verbal reports during a media circus when phrases like "the devil walked in Devon" were being tossed around. Give me something concrete to talk about, and we can look for explanations. But if you must accept these at face value, I submit that most of the animals we've nominated are perfectly capable of fitting inside a four inch drainpipe, and often do enter such structures for security or while foraging. I've already dismissed the river crossing to my satisfaction. And as far as materializing through walls and haystacks, I don't have enough information to speculate. Was there sheltered ground around the base of the wall? Was the wall breached near its end? But mainly, is there really any reason to believe that this happened?

So who did walk in Devon that night? In the presence of so many possible and reasonable culprits, and in the extreme unlikelihood that this was indeed one single set of prints, I find little reason to turn to supernatural explanations. Commonplace events are frequently blown out of proportion, and everyday objects are just as frequently regarded as supernatural oddities. In my opinion, which I think is well supported with other examples in recent history, the devil's walk through Devon was the 1855 version of the Virgin Mary appearing in the bark of a tree or in bathtub stains. People see what they want to see and they think what they want to think, even when they're looking no farther than their own grilled-cheese sandwich skillet.

32. Blood for Oil

The War on Terror and anti-American sentiment in the Middle East has raised our gasoline prices to an all-time high. Or has it?

Everyone knows that most of our oil comes from the Middle East, which is why we're so heavily dependent upon them for our energy. And when they don't like us, for example when we bomb them, they jack up our prices to hit us where it really hurts. Or so I've always heard. But is any of that really true?

More than a third of our petroleum, about 37% of our total usage, is produced domestically by our own oil companies. I'm not sure why people seem to forget about those guys, Exxon-Mobile and Chevron and all of them; you may resent them but they are the principal source of our non-foreign-dependent energy. So this means that only a bit less than two thirds of our petroleum is imported. That still means most of our petroleum comes from OPEC, right? Wrong. Most of our petroleum imports come from non-OPEC countries; 56% of it, in fact. Of that 56%, the majority is from Canada and Mexico, who are about as far removed from the Middle East as can be. The rest of it is from other random places like Angola, Russia, the Virgin Islands, and Brazil, all of whom are friends of ours. So exactly where is all this leverage from anti-US exporters coming from?

You probably assume that it comes from the OPEC countries, who provide the remaining 32% of our petroleum. Well, here's the next monkey wrench. Of that 32%, almost none comes from hostile Middle East countries. The biggest supplier is Saudi Arabia, a relatively Westernized country that's our biggest ally in the region. Number 2 is Venezuela (in more ways than one); their president may be headed for a rubber room but they're hardly a Middle Eastern terrorist nation. Number 3 is Nigeria, and I'll bet you didn't know that they had an industry other than sending emails promising millions. Number 4 is Algeria, and what's left comes from Iraq, which isn't allowed to hate us any more now that we occupy them. In fact, less than 1% of our petroleum comes from hostile Middle East countries.

No blood for oil, say the anti-war protesters. I'm against the war too, but I'm interested to hear that particular claim defended. Yes, the United States does launch some pretty unpopular military actions in this world, but not against anyone who provides any significant part of our oil. *No blood for oil.* Looks great on an anti-war protest sign. Sounds great on 60 Minutes. But what's it based on? I don't know.

So I don't understand. Since none of our oil supply is dependent on these Middle Eastern countries we're always fighting with, how come that fighting affects our gas prices? Sounds like a smoking gun to me. Clearly, we wouldn't be fighting them if we weren't getting some oil out of it somewhere, say the conspiracy theorists. Maybe the Saudis are behind it. Maybe attacking Iraq is a way to please Saudi Arabia. Well, if it is, the fighting sure didn't improve our gas prices much. From what I can see, we've gained nothing by attacking Iraq. We certainly haven't won any free oil or earned any favoritism discounts. So why do the conspiracy theorists draw this connection? I don't know. None of it makes any sense to me.

Yet, something has driven up the gas prices. Whose word do we accept unconditionally: the government's, or that of the anti-US conspiracy theorists? Maybe a liberal dose of skepticism is due. Maybe all of these people are speaking with an agenda, rather than with responsible critical analysis.

An advertisement in the New Yorker magazine costs a lot more than an advertisement in People Magazine, despite the fact that People Magazine reaches many more readers. A Porsche Turbo costs almost twice what a Porsche Carrera costs, even though they're 98% identical. A Rolex contains the same parts as a Timex but costs a hundred times as much. Has the world gone mad? When did prices suddenly jump off the sanity wagon? Since when do companies charge a penny more for their products than their production cost?

Prices are driven by markets. Markets are driven by human beings. Human beings are driven by emotions. Emotions, like fear, explode when we get into a war. When we get into a war in an oil-producing region, petroleum markets in those regions go insane. Stockholders get nervous. Traders freak out. Prices climb the tree like mad to escape the tsunami. Everyone between you and the guy who connects the hose to the well in Yemen becomes terrified, and oil becomes the most prized commodity on the planet. It's simple, it's obvious, it's organic, and it's Economics 101. It's not *blood for oil* and it's not a Halliburton conspiracy. It's a fact of world economics, and the tidal force of the world economy is the strongest superpower on Earth: greater than Dick Cheney, greater than Osama bin Laden, greater than anti-war protesters. The US administration *wishes* it could control oil prices like this.

I'm well aware that this little outburst of mine is not going to change the mind of anyone who believes that the war in the middle east is all about oil. I know that plenty of readers are going to find fault with my research and point out that we did in fact get seven barrels of oil from a hostile country ten years ago. I know that many readers are going to drag out the tired old adage that if Iraq produced asparagus instead of oil, the Gulf War would never have happened. I know that many readers are going to point out that regardless of short-term price hikes, it's essential for the US long-term energy strategy to have a strong military presence in the Middle East. I know I can't change your mind. What I can do is to encourage you to be skeptical for a moment. What I can do is to encourage you to look up, on your own, where our oil actually comes from. When you see how little of it comes from the Middle East, and especially how almost none of it comes from the hostile parts of the Middle East, I hope that you will at least re-examine the *blood for oil* claim. Is that small percentage of oil truly more important than virtually everything else about our country, to the point where we'd infuriate everyone in the world, plus most of our own people, to wage a war? I'm not a politician and I don't claim to know what the war is really about, but when I look at the oil question skeptically, it just doesn't emerge as a logical cause. I'm not claiming to have the answers and I'm not even claiming to be right, but I am claiming to have thought about it more, and personally done more independent research into the sources of our petroleum, than most people who simply parrot the *blood for oil* slogan because it's a great sound bite and because it's an easy and trendy way to be anti-Bush. I'll give you a great starting point for your own research, an article by the engineering editor of Road & Track, at http:// www.roadandtrack.com/article.asp? section_id=18&article_id=3434.

I fully expect this chapter to be among the least popular, and the most criticized. It's always more popular to be skeptical of the government, than to be skeptical of those who *doubt* the government.

33. Homeopathy: Pure Water Or Pure Nonsense?

Let us take a tiny sugar pill, infused with specially charged water, and cure our ills in a novel way. For this chapter's topic is homeopathy, one of my favorite of the many popular alternative medicine systems. Homeopathy has a large following, but I suspect that a large number of its customers don't really understand what it is. For example, I asked two friends who are homeopathy users, on separate occasions, to tell me about it. By coincidence both were attempting to treat headaches. Both friends had the same general understanding of what homeopathy is: They said it was essentially an herbal remedy, and that the small pills they were taking contained some sort of herbal extract. They could not have been more wrong. I wonder if they would continue taking it if they knew what homeopathy really is.

Samuel Hahnemann was a German physician. In the late 1700's, all medical conditions were believed to be caused by an imbalance in the four basic bodily humors: blood, phlegm, black bile, and yellow bile. Conventional medical practice was to attempt to equalize these humors by such practices as bloodletting, purging, or leeching. Hahnemann observed that these practices often caused more problems than they solved, and so he set about developing a better, safer way to balance the four humors. He reasoned that the body may be able to balance its own humors, if given a sort of "kick start" by administering a small dose of whatever poison or toxin was thought to cause the imbalance. He called this the Law of Similars. The obvious problem was that administering poisons and toxins would kill the patient, so he devised a system of massively diluting the ingredients with water. Hahnemann claimed that greater dilutions had greater effect in balancing bodily humors, and he called this the Law of Infinitesimals. His dilutions were as high as 1 part in 10^{30}. This proportion is vastly larger than one grain of sand in all the deserts and all the beaches and oceans on the Earth. He published his theory in 1807, and homeopathy was born.

And then Hahnemann did a very subtle, a very clever, little thing. He made up a word. The word he invented is *allopathy*. Allopathy is Hahnemann's name for all evidence-based medical sciences. That's right: every medical discipline you've ever heard of — including internal medicine, oncology, neurology, cardiology, psychiatry, pathology, surgery, infectious disease, hematology, geriatrics, gastroenterology, ophthalmology, radiology, orthopedics, nephrology, urology, pharmacology, emergency medicine and critical care — they're all simply allopathy. Allopathy is only one word, so it's no better than homeopathy. They're equals. You have the musings of one guy 200 years ago on one hand, and on the other you have everything medical science has taught hundreds of thousands of researchers since then. Homeopathy vs. allopathy. It's nice to be able to conveniently dismiss so much with just one word. This makes it possible to offer the innocent patient Door A or Door B. Knowing nothing further about either choice beyond its one-word name, the innocent victim will probably take whichever the practitioner advises.

Homeopathy shares one very important component with most other alternative medicine systems. It was developed a long time ago, by one man, during a time when almost nothing useful or true was known about medicine, and it is rigidly required to stay frozen in time with the same original ancient world view. Homeopathy, like other alternative medicine systems, does not, cannot, must not grow, evolve, or improve as we learn more about the human body. If it did adapt to new knowledge, it would cease to be homeopathy and would be something different.

This ability to include and adapt to new knowledge is the central strength of modern medicine. When we learn new things about the body, when we find a better way to treat a condition, we adapt. We publish the results and we train doctors on the new techniques. Every day, the knowledge base that modern medicine is built upon grows. The collective experience of researchers and doctors grows. But for homeopathy, and other alternative medicine systems, the knowledge base stays frozen in 1807. AIDS drugs, for example, are so much better now than they were just ten years ago, and ten years from now, they'll be even better (there may even be a cure). But with homeopathy, AIDS is treated the same way that any unknown illness was treated in 1807: with a vial of water, possibly containing a few molecules of some compound that are hoped might stimulate a balance of bodily humors.

Dilutions of homeopathic products that are sold today usually range from 6X to 30X. This is homeopathy's system for measuring the dilution, and it doesn't mean 1 part in 6 or 1 part in 30. X represents the roman numeral 10. A 6X dilution means one part in 106, or one in one million. A 30X dilution means one part in 10^{30}, or one followed by 30 zeros. A few products are even marketed using the C scale, roman numeral 100. 30C is 100^{30}. That's a staggering number; it's 1 followed by 60 zeros, about the number of atoms in our galaxy. In 1807, they knew more about mathematics and chemistry than they did about medicine, and it was known that there is a maximum dilution possible in chemistry. Some decades later it was learned that this proportion is related to Avogadro's constant, about 6×10^{23}. Beyond this limit, where many of Hahnemann's dilutions lay, they are in fact no longer dilutions but are chemically considered to be pure water. So Hahnemann designed a workaround. Hahnemann thought that if a solution was agitated enough, the water would retain a spiritual imprint of the original substance, and could then be diluted without limit. The water is often added to sugar pills for remedies designed to be taken in a pill form. So when you buy homeopathic pills sold today, you're actually buying sugar, water, or alcohol that's "channeling" (for lack of a better term) some described substance. The substance itself no longer remains, except for a few millionth-part molecules in the lowest dilutions.

Let's look again at Avogadro's number. 6×10^{23} atoms is called a mole, a term any chemistry student is familiar with. How big is that number? Well, if you had 500 sheets of paper, you'd have a stack about two and a half inches high, like a ream that you'd buy at the stationery store. If you had 6×10^{23} sheets of paper, your stack would reach all the way from the Earth to the Sun. And not only that: it would reach that distance four hundred *million* times. Think about that for a moment. One sheet of paper, in a stack that's 400,000,000 times the distance from the Earth to the Sun. That's a typical homeopathic dilution. Sounds pretty potent, doesn't it?

One explanation offered by some homeopaths is that it works the same way as a vaccine: putting a tiny amount of a disease-causing agent into the body — not enough to cause the disease, but enough to stimulate the body's natural defenses into fighting off that disease. Well, this is indeed the way a vaccine works, but it's got nothing to do with the way Hahnemann defined homeopathy. Vaccines are used to prevent an illness which does not yet exist in the body by triggering the production of preventive antibodies; and homeopathy is used to fight a disease already in the body, in which case any antibodies would already be in production. The number of the antibodies triggered by a vaccine can be measured in the bloodstream, whereas homeopathy is not intended to, and does not, produce any measurable reaction. Vaccines insert inert versions of the disease-causing agents into the body, where homeopathic substances are the same as that which causes the disease. Finally and most obviously, vaccines contain a large and fully measurable amount of active ingredient, whereas homeopathic remedies contain no measurable active ingredient. So homeopathy can indeed be said to work just like a vaccine; well, at least, it works just like a spiritual imprint of a vaccine.

So why do so many people claim that it works, and swear by it? Homeopathy has been tested over and over again, and though most studies show its effects to be consistent with the placebo effect, a surprisingly large number of studies do show that homeopathy produces results superior to a placebo. But in every one of these cases, doubts have been raised about the quality of evidence in the studies. According to the National Institutes of Health, "Problems include weaknesses in design and/or reporting, choice of measuring techniques, small numbers of participants, and difficulties in replicating results."

A favorite study of homeopaths is that of the British Medical Journal in 1991, a meta-analysis of 107 controlled trials over a 25 year period. The majority of the studies did show some positive results, and homeopaths stop there. They stop short of the Journal's final conclusion, which was "At the moment the evidence of clinical trials is positive but not sufficient to draw definitive conclusions because most of the trials are of low methodological quality and because of the unknown role of publication bias." If you need the term "publication bias" translated, it means the studies that were actually published were those very few where positive results happened to have been found, rather than the vast majority where nothing or negative results were found. The British Medical Journal went on to say "This indicates that there is a legitimate case for further evaluation of homeopathy, but only by means of well performed trials."

Well, good luck to you, gentlemen. The UK Society of Homeopaths has stated "It has been established beyond doubt that the randomized controlled trial is not a fitting research tool with which to test homeopathy." In other words, homeopathy has given itself a *Get Out of Jail Free* card. Tests are not adequate to test them. If you perform a clinical trial, and find that homeopathy is no more effective than a placebo, the reason for the failure is that homeopathy should not be tested. Claimed immunity from scientific scrutiny should stand out as a huge red flag. When you hear anyone defend their claim by stating that its effect cannot be detected through testing, be skeptical.

The upside of homeopathy is that it's not going to hurt anyone, since it lacks any measurable active ingredients. And when treating conditions that are not life threatening, like headaches or fatigue, there's no harm done. There is massive harm done when practitioners or store owners recommend homeopathy as a replacement for real medical treatment when a serious illness exists. Be vigilant, and protect the health of your family, your friends, and yourself.

34. REVISIONIST DARWINISM: THE THEORY THAT COULDN'T SIT STILL

Now we're going to take a step back from the skeptic's Rock of Gibraltar, evolution, and examine whether it truly has any value as a theory, since we keep having to revise Darwinism.

Darwin's original theory of evolution was generally correct, but it was highly incomplete and has been growing and evolving (pun intended) as we've learned new information since *The Origin of Species* was published in 1859. No evolutionary biologist doubts the fundamentals of evolution, which are essentially as Darwin described them, even though there are numerous minor points that are still under debate or still incomplete. This hardly invalidates the entire theory as a whole. Every significant major point of evolution is proven correct by the evidence. The fact that evolutionary biologists are still employed at their jobs, and still doing research and learning about our world, leads many creationists to use the term "revisionist Darwinism", as if the theory's ability to incorporate new knowledge is a weakness.

Let's go back to a previous chapter, *An Evolution Primer for Creationists,* where we discussed the definition of the word theory. Among the requirements for an idea to qualify as a theory is that it "must allow for changes based on the discovery of new evidence. It must be dynamic, tentative, and correctable." The central strength of the scientific method is that we allow our theories to be improved as new information is discovered. This is why the theory of evolution is a hundred times more rock solid now than it was 150 years ago. This is why modern medicine has doubled the average human lifespan in just the past century. This is why Moore's Law allows us to double the speed of computers every two years. This is why we fly around the world in airplanes. When we learn new information, we accept it, and adapt our theories of the world to accommodate it.

When creationists say "Revisionist Darwinism", they think they're shooting holes in the theory by pointing out that we have to keep revising evolution to accommodate new information, like some worldwide game of whack-a-mole. This is not the way to criticize science. The ability to improve a theory as knowledge improves is the central *strength* of science. Revisionist Darwinism? You're damn right, we revise and improve it every day. That's called doing science.

Pseudosciences and faith-based belief systems, on the other hand, do not accept new information. Let's compare what scientists did and what creationists did in the mid-1900's when DNA was discovered. For evolutionists, the discovery of DNA and the understanding of genetics, unknown in Darwin's time, was a huge windfall. Whole chapters of proposed mechanisms were thrown out of the evolution textbook, volumes of new chapters were added, and unanswered questions were explained by the thousands. The theory of evolution improved immeasurably. Genetics was the single most important discovery in the history of biology. What did creationists do with that information? Did anyone go back and improve Genesis? Did they add a footnote or a verse to explain how the thing with Adam's rib worked, given the new understanding of genetics? No. They did nothing. The most important and significant discovery in the history of biology was completely 100% ignored by the creationists. In creationism, the process of learning is taboo. This explains why when evolutionists embraced genetics, creationists saw it as a weakness and they made up condescending terms like "revisionist Darwinism".

There is another irony that's all over creationist arguments against evolution. One thing we hear a lot is that to accept evolution, you have to *believe* in the fossil record, you have to *believe* in what other scientists tell you, you have to *believe* what radiocarbon dating machines reveal — in short, that evolution is all about belief, and that it's therefore faith-based. In fact, that Darwinism is just another faith-based religion. That's my favorite irony: creationists criticizing evolution by calling it a faith-based religion. And creationism is ... what again? However, this comparison is completely invalid. Creationists are confusing faith with trust. When we accept or believe the results of a radiocarbon dating test, or when we accept and believe the fossil record, that's trust, not faith. Trust is when you accept what well-sourced evidence tells you. Faith is believing in something despite evidence to the contrary. When I compute a figure on my RPN calculator, I don't have faith in the answer. I trust the answer. I have good reason to accept that answer as fact. Calculators have been shown to be reliable to so many 9's that it's hardly worth mentioning. Now, if my calculator was usually wrong or if the results of every computation were of unknown validity, then faith would be the right description for my acceptance of the answer. Trust is when you accept what well-sourced evidence tells you. Faith is believing in something despite evidence to the contrary. Evolution is not a faith-based religion, but nice try, fellas.

Previously we've talked about homeopathy and reflexology in this book. Both of these systems were developed many decades ago, before the advent of modern medicine, when almost nothing useful or true was known about the human body. The state of medical science sucked. Paranormal explanations were proposed by the founders of homeopathy and reflexology, in honest attempts to understand medicine during a time when no good information was available yet. In all the years since, many of the true workings of the human body have been learned. Medical science adapted and improved. Call it "Revisionist Medicine" if you like. Modern medicine is being revised and improved at least as fast as any other branch of science. But not for homeopaths and reflexologists. Their techniques have not accepted what science has learned about the human body; rather, they remain entrenched in the same ancient world views in which they were developed. Homeopathy is still fundamentally about balancing the four basic bodily humors, and reflexology is still about a mystical energy field called "life force" that's centralized in your feet and hands. To accept homeopathy or reflexology, you must believe in it despite evidence to the contrary. That's faith. To believe in modern medicine, you need only accept what well-sourced evidence tell us. That's science.

History is full of examples of mankind's collective knowledge improving. We used to think the world was flat, and was supported on the backs of giant tortoises. Later we thought that the Earth was the center of the universe. Some people accept what we learned through scientific research, and some people don't. Most people accept some of it and deny some of it. Some people say that since we used to think the world was supported by tortoises, all of modern cosmology and astronomy must be invalid. It's a logical fallacy, but that's what a few people believe nevertheless. By the exact same logic, some creationists consider evolution invalid because biologists employ learning to improve their knowledge. It's another logical fallacy, but again, everyone's entitled to their opinion. Some are just more correct than others.

35. Mercury Fillings: A Mouthful of Death!

When you go to the dentist and get a tooth drilled out, chances are he's going to fill the hole with an amalgam filling, which has been the gold standard of dentistry for more than 150 years. Amalgam fillings consist mostly of silver, tin, and mercury, in that order by volume, with some other stuff mixed in. Mercury was used in gold and silver mining because it readily absorbs those elements, keeping them securely bound chemically. Amalgam is the filling of choice of nearly all practicing dentists because they're cheap, durable, easy to work with, and generally considered safe. So safe, in fact, that the American Dental Association has never reversed that position in 150 years. It is acknowledged that amalgam fillings do release molecules of all constituent metals, but not in significant amounts. Wristwatches, rings, and other jewelry have been shown to introduce their surface metals into the body at at least the same rate, but as we know from the last several thousand years, this has never been a problem for anybody. We're talking about infinitesimal levels.

Therein lies the disputed point. A small but very vocal minority of dentists and other people believe that the levels of mercury released from amalgam fillings is high, and in some cases that it's dangerous or even lethal.

I am personally acquainted with two people who suffered for years from chronic fatigue, and both reported immediate relief when they had their amalgam fillings removed from their teeth and replaced with ceramic fillings. It should be noted that the true causes of Chronic Fatigue Syndrome are well understood to have nothing to do with heavy metal poisoning. Hal Huggins, DDS is a controversial author who has written a number of books warning about the dangers of mercury from amalgam fillings. Read the reviews of his books on Amazon, and you'll see that he has a huge following of believers singing

Halleluias to him. Clearly, the belief that amalgam fillings cause toxic mercury poisoning is widespread and popular. Correspondingly, the belief that removal of amalgam fillings causes an immediate reversal of all symptoms is equally widespread.

Now, these people may indeed have experienced immediate relief upon having the procedure done — I know my two friends did. But I'm doubtful that their relief was not at least partially psychological. The human body has no mechanism for removing heavy metals from itself. This can only be done through chelation therapy, a lengthy and tedious process (and it's important to distinguish real chelation from the popular pseudoscience version used for "cleansing"). Replacing the fillings may have stopped the alleged introduction of new mercury, but it would not reduce any existing mercury levels in the body. If mercury poisoning is indeed the cause of the patient's problem, removal of the amalgam fillings would not produce any relief at all.

And, of course, nobody is exposed to more amalgam fillings than dental professionals. The lack of any increased level of mercury poisoning among dental professionals tends to cast additional doubt onto the claim that amalgam fillings represent a significant source of mercury poisoning.

The most compelling evidence of the dangers of mercury leeching from amalgam fillings is the so-called "Smoking Teeth" video. You can see it if you search YouTube for "smoking teeth" or just go to iaomt.org. IAOMT is a group of dentists and other scientists who believe that amalgam fillings release toxic levels of mercury into the body, and they call themselves the International Academy of Oral Medicine and Toxicology. Their video shows a recently extracted tooth that contains an amalgam filling, and which has been dipped in water, held in front of a fluorescent screen, and illuminated with ultraviolet light. This is an easy way to show visually whatever vapors might be coming off of the tooth. Well, the video shows a constant dense smoke of vapor rising from the tooth. The video begins with the direct claim "All mercury silver fillings leak substantial amounts of mercury constantly. The amount increases with any kind of stimulation." The narrator proceeds to rub and heat the tooth, and the visible smoke rising from the tooth increases accordingly. The video is quite alarming, and even I felt a wake-up call when watching it.

But then it occurred to me: Isn't mercury a lot *heavier* than air? Wouldn't mercury vapor drop to the floor like a rock, or like CO_2 mist from dry ice? Why would it be *rising* from the tooth? I even double checked my periodic table to be sure. I did a little bit of research on the web to see what I could find out. And, sure enough, everything I found confirmed my suspicion. Mercury vapor is much, much heavier than air. Whatever's rising from that tooth in the video can't possibly be mercury vapor. Discussing this with a friend, I learned that a simple yet thorough debunking of this video has already been done, by Dr. James Laidler, MD, and you can find his short but very clear article on his blog, at quackfiles.blogspot.com. If you doubt anything in Dr. Laidler's article, you can quickly glance at any periodic table of the elements and confirm it. The simple fact is that at body temperature, air weighs 1.2 grams per liter. Mercury vapor weighs 7.86 grams per liter, more than six times heavier than air. The vapor in the video is rising, fast enough to indicate that it weighs — oh, around .71 grams per liter. And guess what weighs that much? Water vapor. Remember they said on the video the tooth had just been dipped in water? That's right: This shocking video, the centerpiece of evidence in the case against amalgam fillings, points directly to a column of rising water vapor and tells you that it's mercury vapor, in direct contradiction to chemical fact. The credits on the video are from two IAOMT dentists, Roger Eichman, DDS and David Kennedy, DDS. According to an IAOMT representative, the video was made by Boyd Haley, Ph.D., a professor at the University of Kentucky. And guess what he's a professor of? Wait for it: Chemistry.

The same IAOMT representative told me: "No one with any knowledge of these subjects could ever say that the video is quackery as these guys know their stuff and have presented to Congress on multiple times, and have used this video extensively in their campaign. THE VIDEO IS AN

IRREFUTABLE PIECE OF SCIENTIFIC FACT - PERIOD!" It appears that "irrefutable" has a new definition. Apparently now, it means "completely wrong and deliberately misleading." Unless he's asleep all day, you've got to figure that Professor Haley knows that mercury vapor is far heavier than air, and would sink. I see that this comes down to three possibilities. (1) He's honestly wrong about something this flagrant, which seems unlikely; (2) He is deliberately lying in the video, which seems unlikely; or (3) He explains his position with some excuse like there is mercury vapor mixed in with the water you see rising. All right, let's assume it's number 3. Any mercury vapor would still fall; mercury vapor would not remain in solution with water vapor. So whatever vapor we see rising has nothing to do with the mercury amalgam filling said to be in the tooth in the video. Any tooth dipped in water would smoke in exactly the same way.

Now here's where I get confused. These guys are not selling anything. All indications are that they're genuinely concerned with people's health. They're educated professionals. Their hypothesis is a fringe idea, but that doesn't make it wrong. New discoveries often start on the fringe. But if all that's true, why did they make this video, which is full from beginning to end with flagrant errors and poorly presented evidence?

Now let me be very clear on one point: I'm not here to say that amalgam fillings are safe. I'm not making any claims about that one way or the other. What I'm saying is that these IAOMT people are supporting their claim badly, with bad science. The claim that amalgam fillings are unsafe may or may not be true, but the claimants have yet to produce any real evidence (that I've seen) that passes any close scrutiny. Just as another couple data points: IAOMT opposes the use of fluoride, and they also say 85% of dentists are impaired by mercury poisoning. My dentists have never heard of anyone. I guess they must be too impaired to be able to tell. IAOMT might be right about all of this stuff; they're just not making a very good show of it.

I will go on the record right now with my personal opinion, whether you care what that is or not: I don't want anything with mercury going into my mouth when there are alternatives, especially when those alternatives look better cosmetically. In 2006 the FDA did overrule its previous position that amalgam fillings are safe, and say that more study is needed. That's not saying that they're unsafe, it's simply saying that more study is needed. I've had amalgam fillings in my mouth my whole life — I even remember playing with the raw mercury on a countertop when my dentist uncle put some of my earliest fillings in, which obviously isn't too good — and I don't have any mercury poisoning. I'm not about to get my fillings replaced, but I do choose ceramic fillings now; but mainly because they look better.

If you're among those considering having your amalgam fillings removed, don't take any unnecessary medical procedure lightly. Consult not only with fringe books and videos you find on the Internet, but also consult with your dentist and your family doctor. Also consider the greater surface area contact of wristwatches, rings, and other jewelry if you truly believe yourself to have heavy metal poisoning. Nobody's ever been proven to have significant levels of mercury in their body from amalgam fillings, so don't worry too much about it regardless.

36. How to Spot Pseudoscience

Carl Sagan was one of the first to bring to the mass public a toolkit for deciphering science from pseudoscience when he published *Demon-Haunted World* in 1996. His "baloney detection kit" was a set of guidelines to help lay people spot logical fallacies and other common flaws in popular pseudoscientific claims. Many other great minds have continued in this tradition, and today we're flush with similar lists. In an effort to increase the confusion, I've taken it upon myself to compile some of the best, mix them all up, pick and choose the best points, and distill them down into a single list. In compiling this list, I owe the biggest debts to the brilliant Dr. Stephen Barrett of Quackwatch.org, Dr. Tom Perls of AntiAgingQuackery.com, Dr. Michael Shermer of the Skeptics Society, and of course Dr. Carl Sagan. All four of these guys have put in many years of hard work protecting the innocent public from harmful and untruthful pseudoscientific claims. It's no accident that common themes run through all of their work.

I've created a 15-point checklist that I call "How to Spot Pseudoscience." When you hear any claim about a new product, a new discovery, or some paranormal ability, run it through these fifteen questions and you'll get a pretty clear idea of whether or not it has any merit.

1. Does the claim meet the qualifications of a theory?

Very few claims that aren't true actually qualify as theories. Let's review the four main requirements that a theory must fulfill. 1) A theory must originate from, and be well supported by, experimental evidence. Anecdotal or unsubstantiated reports don't qualify. It must be supported by many strands of evidence, and not just a single foundation. You'll find that most pseudoscience is supported by only a single foundation. 2) A theory must be specific enough to be falsifiable by testing. If it cannot be tested or refuted, it can't qualify as a theory. And if something is truly testable, others must be able to repeat the tests and get the same results. You'll find that this feature is truly rare among pseudosciences; they'll generally claim some excuse or make up a reason why it can't be tested or repeated by others. 3) A theory must make specific, testable predictions about things not yet observed. 4) A theory must allow for changes based on the discovery of new evidence. It must be dynamic, tentative, and correctable. You'll find that most pseudoscience does not allow for changes based on new discoveries.

2. Is the claim said to be based on ancient knowledge?

This is a sure sign that the claim is not based on scientific evidence, and it's intended to fool you into thinking that because the ancient Chinese believed it, it must have merit. In fact many true theories are not very old at all, because they've replaced older theories as knowledge has increased. Generally, the more recent the evidence, the better scientific foundation it has.

3. Was the claim first announced through mass media, or through scientific channels?

Real discoveries go through an unbiased peer review process, which results in publication through scientific journals. When a belief is first announced through the mass media, like Pons and Fleischman's cold fusion experiments or like the Stoern perpetual motion machine, there's generally a reason its proponents chose not to subject it to the scrutiny of peer review.

4. Is the claim based on the existence of an unknown form of "energy" or other paranormal phenomenon?

Loose, meaningless usage of a scientific-sounding word like "energy" is one of the most common red flags you'll see on popular pseudoscience. Terms like energy fields, negative energy, chi, orgone, aura, psi, and trans-dimensional energy are utterly meaningless in any scientific context. Approach with extreme caution.

5. Do the claimants state that their claim is being suppressed by authorities?

This is usually a really frail excuse for why mainstream scientists don't take their claim seriously, why the product is not approved by the FDA, or why scientific journals won't publish their articles. You'll often hear this in the form of a conspiracy of the medical establishment to suppress a quack cure because it's in the interest of the medical industry to keep you sick. In fact, any doctor or pharmaceutical company that could develop a new cure would make a huge fortune; they'd never suppress it. The same goes for auto manufacturers worldwide who are said to be "suppressing" new efficient engine technologies.

6. Does the claim sound far fetched, or too good to be true?

When something sounds too good to be true, it usually is. Extraordinary claims require extraordinary evidence. Does the claim truly fit in with what we know of the way the world works? How often do claims that turn the world upside down really turn out to be true? Approach such claims with extreme skepticism, and demand evidence that's as extraordinary as the claim.

7. Is the claim supported by hokey marketing?

Be wary of marketing gimmicks, and keep in mind that marketing gimmicks are, by themselves, completely worthless. Examples of hokey marketing that should always raise a red flag are pictures of people wearing white lab coats, celebrity endorsements, anecdotes and testimonials from any source, and mentions of certifications, colleges, academies, and institutes.

8. Does the claim pass the Occam's Razor test?

Is there a simpler, natural explanation for the claim that does not require any supernatural component? Are results consistent with the placebo effect or the body's natural healing capacity? Can a stage magician duplicate the psychic's feats? The Law of Large Numbers states that a one-in-a-million event usually happens to everyone about once a month, and since Occam's Razor says that the simpler of two possible explanations is usually the right one, don't leap for a supernatural explanation just because you happened to dream about your grandmother on the night she died.

9. Does the claim come from a source dedicated to supporting it?

Science works by starting with a null hypothesis and searching for evidence. Pseudoscience starts with a positive hypothesis and supports it with questionable research and anecdotal reasoning. It's unlikely that an institution dedicated to the promotion of any given claim will present any type of evidence other than that which supports their claim, and its bias should be given serious consideration.

10. Are the claimants up front about their testing?

Any good research will outline the testing that was done, and will present all evidence that did not support the conclusion. Be skeptical of any claims that do not detail testing methodology that was thorough and responsible, including external verification and duplication, or that do not provide evidence unsupportive of the conclusion.

11. How good is the quality of data supporting the claim?

Watch out when testing data might be susceptible to observational selection, which is the counting of hits and not the misses, like we see with television psychics. Watch out when sample sizes are too small to have statistical significance, as with most clinical trials of homeopathy. And especially watch out for hastily drawn causal relationships: the assumption that because the relief occurred after the remedy, the remedy must have caused the relief.

12. Do the claimants have legitimate credentials?

Be aware that there is a huge number of unaccredited institutions (which are often just bedroom offices) giving out degrees in just about anything. Be aware that some institutions claiming to be accredited received their accreditation from unrecognized accreditation bodies. Finally, be aware that genuine accredited universities often have programs in unscientific fields such as chiropractic, naturopathy, and acupuncture. You must be vigilant. To see just how vigilant, go to ThunderwoodCollege.com and get your own Ph.D. in the field of your choice in seconds, for free.

13. Do the claimants state that there's something wrong with the norm?

When real research is presented, it consists of the evidence that was discovered and the conclusion. It does not go off on alarmist rants about how the food we eat is dangerous, how we're destroying the planet, how the government covers up its evils, or how you're going to hell if you accept evolution. When a claim is presented as an alternative to the wrongs of the status quo, it's a sign that the claim is probably based on ideology or philosophy rather than science.

14. Is the claim said to be "all natural"?

As we've see time and time again, by no definition can "all natural" mean that a product is safe or healthy. Consider the examples of hemlock, mercury, lead, toadstools, box jellyfish neurotoxin, asbestos — not to mention a nearly infinite number of toxic bacteria and viruses *(E. coli*, salmonella, bubonic plague, smallpox). In many cases, synthetic versions of natural compounds have been engineered to make them safer, more effective, and able to be produced in large quantities.

15. Does the claim have support that is political, ideological, or cultural?

Some claimants suggest that it's moral, ethical, or politically correct to accept their claims, to redirect your attention from the fact that they may not be scientifically sound. In some cases, such as creationism, proponents use the court system to force schools to teach their claims as fact. Generally, when a theory is scientifically sound, even if it's brand new it will eventually find its way into the educational curriculum. Good science is done in the lab — not in the courts, not in protest marches, not in blogs, and not in church. A political or cultural campaign to legalize or promote some product or claim is a major indicator that it's bogus.

There you have it. With this checklist, anyone is well equipped to filter out the chaff from the wheat. Questions like these are what should be taught in schools, encouraging young people to begin looking at all the crazy misinformation in our world with critical analysis. The ability to tell fact from fiction is essential to our progress as a species as we search for the next great discoveries in medicine, space exploration, computing, power generation, and every other scientific field.

37. THE MARFA LIGHTS: A REAL AMERICAN MYSTERY

Tonight we're going to zip up our windbreakers and camp out in our folding chairs, drinking coffee from a thermos, until the ghostly Marfa Lights make their appearance, hovering and wavering out in the field before us.

In 1957, a magazine article first reported the mysterious phenomenon of hovering balls of light bouncing around the night near Marfa, Texas. About ten miles east of Marfa, in an area called Mitchell Flat, the odd lights appear perhaps a couple dozen times per year. They're about the size of basketballs and appear to float about shoulder high. Sometimes several appear at once, wavering about, sometimes even merging together or moving about in a group, splitting apart, and behaving in a most remarkable fashion. They only appear at night, at any time of year and in any weather, and are usually white or yellow or orange. Sometimes red or blue are reported, but most are white.

One unique characteristic of the Marfa Lights is that they are actually there and can actually be observed; their existence is definitely not just a story. The city of Marfa has even erected a viewing platform, where hopeful light spotters can be found every night. You can actually go there, and if you're lucky or patient enough, you will actually see the Marfa Lights.

The default skeptical explanation which is readily put forth is that the Marfa Lights are simply car headlights, seen from a great distance and distorted by temperature gradients. Critics of this explanation quickly point out that the Marfa Lights have been reported for hundreds of years, since long before there were any car headlights around.

Well, apparently, the Marfa Lights have *not* been around all that long, after all. The earliest accounts come from a rancher named Robert Ellison in 1883. However, upon closer inspection, it appears that there is no actual record that Robert

Ellison ever saw such a thing. There are reports from his descendants that Ellison *said* he saw lights, but there is no written record, not even when he wrote his memoirs about his life in the region in 1937. Curious that he would leave that out. Apparently, all evidence that the lights existed prior to the arrival of automobile highways in the region is purely anecdotal. Throughout history there have been hundreds and hundreds of reported "ghost lights" that probably never existed outside of the observers' whiskey-soaked imaginations. Those ghost lights that have become famous are those few that are observable today, such as the Marfa Lights, and almost all have multiple versions of illustrious histories invented for them. A similar phenomenon in Arkansas called the Gurdon Light is said to be the swinging lantern of a brakeman accidentally beheaded by a passing train. Not surprisingly, the exact same explanation is put forth for the Big Thicket Ghost Light in Bragg, Texas. These ghost lights can't all be headless brakemen, so it's conceivable that the folk explanation is not true in every case.

So what are these lights, you ask, and why can't someone run out there and track one down? There are two reasons why this is easier said than done. First, it's illegal. All the property in the area is privately owned, most of it by people who are tired of these tourists chasing around in the night, and do not welcome them. Second, the terrain is exceptionally difficult to traverse. However, hardy souls have made the attempt on numerous occasions. At least two television shows have done just this, and wisely staked out people in different locations in an effort to triangulate and precisely locate the lights. But, as luck would have it, these triangulation efforts have never been successful. The only thing conclusively demonstrated was that the lights are not where they appear to be. This has effectively put to bed other theories, such as the suggestion that piezoelectric effects from the quartz bedrock causes ball lightning. If ball lightning was floating around out there, searchers would be easily able to triangulate and close in on it.

Unfortunately for all of those like myself who love a good mystery, and contrary to what's printed in all the Marfa tourism brochures and on all the ghost light web sites, the Marfa Lights have been thoroughly and definitively explained. The same explanation most likely applies to many similar lights around the world. If you prefer a mystery and don't want to hear it, then stop reading now. It's a spoiler, and like all spoilers, it's disappointing.

May I have the envelope, please? The winner is ... the car headlights combined with some fascinating atmospheric phenomena. In 2004, The University of Texas sent the Society of Physics Students, a highly respected professional association, to investigate the Marfa Lights. Their official report, available at spsnational.org, found conclusively that when the lights appeared, they were precisely correlated with car headlights on Highway 67. The lights were completely predictable and the phenomenon was fully repeatable, based on cars on the highway. Quite a few photographs have been taken of the lights at night, which when superimposed upon a photograph from the same camera location during the day, show Highway 67 in the extreme distance, precisely in the same place as the light in the night photograph. The strange movement of the lights is attributed to the magnifying or shimmering effect caused by a so-called Fata Morgana mirage, a type of superior mirage, in comparison to the more common inferior mirage. Superior mirages, where objects appear higher than their actual position, can make distant objects — even those below the horizon — appear to hover in the air. Inferior mirages, where objects appear below their actual position, can make objects up in the air, such as a patch of sky, appear below the horizon, like the proverbial lake in the desert. Anytime the temperature gradients are suitable, the Marfa Lights should appear and behave predictably. Other independent investigations have also found the same correlation with cars on nearby Highway 90.

There are still critics who do not accept what the investigations have revealed, in some cases because of the stories of reports that predated the highways. These critics are reminded that there is no record of Marfa Lights reports before the appearance of automobiles in the area. And anyway, a lantern or other light would be affected in exactly the same way as the car headlights are today.

Some reports talk of the lights hovering above a person, moving around them, or behaving in other ways inconsistent with the SPS explanation. The Marfa Lights have been photographed and videotaped exhaustively, but none of that evidence supports such reports. Many such reports are probably hoaxes, imagination, or exaggerated recollections. However, it is very difficult to judge the distance of a light source distorted by a Fata Morgana inversion layer. Many reports of people getting within a few meters of a Marfa Light are probably quite genuine: Every visible indication would be that the light appears to be hovering eerily just out of reach.

Considering the evidential consistency and comprehensiveness of the SPS investigation, and its numerous independent verifications, I see no reason to doubt that the mystery has been conclusively solved, and certainly see no need for bizarre and unprecedented alternative explanations like the piezoelectric lightning balls. I'm still a little worried about encountering a headless lantern-swinging brakeman in the dead of night, but that's only because I'm afraid he might want *my* head, Ichabod Crane style.

38. Heating Up to Global Warming

There are obvious signs that everyone can see, and that aren't being debated: the Earth is on a warming trend over the past century or so. Ice shelves and glaciers have been shrinking alarmingly everywhere. We know that there's more carbon dioxide in the atmosphere than ever before. Almost all climatologists draw a causal relationship between temperature and CO_2, saying that CO_2 creates the greenhouse effect, thus causing the higher temperatures. A few people draw the causal relationship in the other direction. Most people say the current trend is truly remarkable, but only if you choose a recent segment of climate history. Some people say that if you look back far enough you'll find many such ups and downs that have been far more dramatic than this one and have lasted many times longer, before there were those evil pesky humans around to cause it. The big question, which is significant because it's one that we probably can do something about, asks how much is human activity to blame for the current trend. Everyone agrees that it's at least partly to blame, but the estimates of how much range from 100% to .00001%. Regardless, it's non-zero, and it really should be zero, and everyone agrees that making it zero should be a goal. But we're left with many intractable questions: How much can we do? How much should we do? How much do we need to do? How much can we afford to do? How much can we get away with? All things considered, where do lines really need to be drawn?

Lots of people pretend to have the answers to those questions. In any debate on global warming, both sides will generally say something like "When you actually look at the data, it shows X, not Y." Well, who actually *has* looked at the data? I don't pretend that I have. I haven't gone down into the cave myself and examined the ice cores under the gas chromatograph. I haven't looked up the raw data from ocean temperature measurements in the Aegean Sea. I haven't personally measured rainfall across southern Africa over the past 100 years. I've seen Al Gore's movie, but that's a meta-analysis of some guy's meta analysis of some other guy's meta analysis. None of the clowns out there who presume to speak authoritatively about what the actual data shows have looked at any raw data. They've looked at someone's meta-analysis of data collected from many sources. Please, next time you're having a conversation on global warming, don't tell us what the data actually shows, because you saw a guy on TV tell us what the data actually shows. No one person can or ever will "look at all the data." A person can look at an infinitesimally small chunk of data that's out of all meaningful context, but let's get real here: the Earth is about the most complicated system imaginable. NEC's Earth Simulator supercomputer, for years the fastest supercomputer on the planet, is dedicated to this task, and they still can't tell us whether it's going to rain tomorrow. Think about that. The Earth is simply way too complicated for any person to be able to claim to understand.

Al Gore says he understands it, and he made the movie *An Inconvenient Truth* to tell us how alarming the situation really is. Senator James Inhofe says he understands it, and he wrote the *Skeptic's Guide to Debunking Global Warming* to tell us how alarmed we should be at how alarmingly the alarmists alarm us. Whether you're feeling alarmed or not, I'm sure you agree that it's most responsible to listen to both sides if this is an issue where you feel taking sides is appropriate. I'm afraid I don't see it that way. I'm skeptical of anyone who says we know how much we have to reduce emissions.

Here's the way I look at global warming: I don't personally have enough expertise to accurately interpret all the information flying around from both sides, and I can't claim to know for sure how much human activity is responsible for our current trend in average temperatures. But there is one thing that I do know, from simple common sense: Pumping carbon dioxide or any other pollutants into our air is bad. In the United States, the environmental movement successfully killed nuclear power, requiring us to depend almost entirely upon coal, oil, and natural gas power plants. Various studies put the number of annual premature deaths in the United States caused by emissions from these power plants at between 30,000 and 60,000 (thank you, environmental movement). Imagine at least one 9/11-style terrorist attack every month, and that's our ongoing death toll caused just by our power plants. This doesn't even include emissions from manufacturing or transportation. Imagine what this number must be in China, a country with 500 times our annual death toll from coal mining accidents. No matter how you slice it, atmospheric pollution is a horrible, horrible thing. So one way to look at it is — global warming aside — we should stop all emissions immediately, now, yesterday. Everyone already agrees it would be great if we lived in Fantasyland and had zero emissions. And little birds singing on our fingertips.

But it's not as simple as that. As evil and politically incorrect as it may sound, the fact is that everything has its cost/benefit ratio. Let's say we turned off all the coal, oil, and natural gas fired power plants in the country. In about 30 minutes, we're back in the Bronze Age. Everyone on life support in a hospital is dead. Every factory stops production. The Gross National Product drops to virtually zero almost immediately. We run out of food in about a week and start cannibalizing each other. That's not the answer.

Clearly, either extreme is unacceptable, in fact ridiculous. We musn't keep generating greenhouse gases at the current rate, and we can't simply stop it all. And in the attempt to find a happy medium, we can't expect every individual and company to make expensive and complicated changes, in many cases without good alternatives, out of the generosity of their hearts. Sheryl Crow can sing all she wants, but people and industry are still going to do what they need to do. So this means that to get anything done, we have to impose rules and regulations on everybody — like, for example, the Kyoto Protocol, an international agreement attempting to address 55% of global greenhouse gases from 160 countries.

I believe the United States was clearly right in its refusal to ratify the Kyoto Protocol, because of its fundamentally nonsensical exemptions. In short, the Kyoto Protocol restricts nations based on how *wealthy* they are, *not* based on how much greenhouse gas they produce! The United States would have had to adopt economy-strangling restrictions, while China, which will surpass the United States as the world's largest producer of greenhouse gases by 2010 at the speed at which an IndyCar passes a hobo pushing a shopping basket, remains exempt from any restrictions. India, the world's third largest producer of greenhouse gases, is also exempt. Even Al Gore says that 30% of global CO_2 emissions come from forest burning in the exempt third world nations. That's a pretty big chunk that nobody seems to talk much about. Some interpretations have said that without additional controls on the exempt nations, the Kyoto Protocol would result in eventual *increases* in the total greenhouse gas output. By these interpretations, the Kyoto Protocol is merely a symbolic political statement and not a useful tool for reducing greenhouse gases. Personally I think it was simply a case of too many cooks and conflicting interests. Blanket proclamations like the Kyoto Protocol are not the way to approach the problem with any workable practicality. In fact, 13 of the 15 European nations who did ratify Kyoto have been unable to comply with its requirements. It's been a huge failure.

Most reasonable people agree that reducing pollution is generally a good goal, and that it should be done wherever the cost/benefit analysis tips the scale. The costs of making changes can be determined with reasonable accuracy by the pencil pushers and the bean counters. Where these equations become foggy is on the benefit side of the scale. Is the only real potential benefit to be gained the opportunity to have a nice pretty smog-free view of the countryside? Or is the potential benefit our very survival in the face of immediate global catastrophe? How much would you pay for one, and how much would you pay for the other? We simply don't know what we can actually buy here, what we'll actually get for our money. Spend a billion dollars to retrofit factories with carbon dioxide recapturing technologies, and how much is that going to help? Exactly what effect will that have on Al Gore's charts and graphs? Nobody has any idea. It's like we're playing a global game of The Price Is Right: We're standing here with a fistful of dollars, we don't know what's behind any of the doors, and everyone in the audience is shouting.

There is a way to find out what we can actually achieve through the reduction of greenhouse gases, and thus know how much of a reduction we *need* to make, plan a way to pay for it and actually make it happen: Doing more science and learning more about our planet. And we're already doing that. More climatologists are working on the problem than ever before. The risk we face now is getting stuck in the rut of doing nothing until it's too late, waiting for answers that will never fully come.

39. Neanderthals in Present Day Asia

It's time to set aside our sliderules and put on our galoshes, as we prepare to dig through muddy burial sites high in the Himalayan foothills, looking for conclusive proof that Neanderthals survive today, hidden away in the rocky corners of darkest Asia.

I've always loved reading about monsters and ghosts and mysterious creatures. The thing that first caught my intrigue about the Abominable Snowman was that the reports are really about two distinct, and very different, species. The big hairy ape running around the mountains, pursued by Sir Edmund Hillary and leaving the famous footprint photograph, and plaguing thrill seekers in Disneyland's Matterhorn, is usually called the Yeti. There are many variations of this name, but let's stick with Yeti for simplicity. The other species is neither big nor especially hairy nor much of a snowman. The Almas is most often described as a small stout person, about five feet tall, hairy and stinky and mute and living in paleolithic squalor up in the foothills of the Himalayas and into eastern Europe. The description is generally consistent with what we call a Neanderthal. But since there are no remains or genetic evidence, it could also just as plausibly be said to match any of several earlier and more primitive species of the *Homo* genus.

Really quick history lesson on Neanderthals. They are not an ancestor of modern humans. *Homo neanderthalensis* is descended from a separate branch that split off from the evolutionary tree about 516,000 years ago, according to some research published in Nature. Mitochondrial DNA studies have shown conclusively that *Homo neanderthalensis* and *Homo sapiens* evolved separately. As the Cro-Magnon *Homo sapiens* moved in across Europe about 45,000 years ago, *Homo neanderthalensis* was pushed out into little corners of the world. The last known population died out in the vicinity of Gibraltar about 24,000 years ago, thus ending their approximately 300,000 year existence.

This was not the only time the Earth was inhabited by significantly different hominid species. *Paranthropus*, a member of the muscular and apelike robust australopithecines, frequently encountered and competed with the more modern, smaller, and smarter *Homo habilis*.

And so we have the hypothetical existence of the Almas, replete with evolutionary precedent. We have a similar precedent for the Yeti from the fossil record. *Gigantopithecus* was a prehistoric great ape that stood nine feet tall and weighed three times as much as a modern gorilla. Its closest relative living today is the orangutan. *Gigantopithecus* did live in China and India, in fact it lived generally where the Yeti is said to exist now. They first appeared about five million years ago and became extinct only about 100,000 years ago. Like the Neanderthal, they reigned for a long time and only disappeared in the most recent of evolutionary moments. Recall that coelacanths were thought to have gone extinct almost 100 *million* years ago, and they gave us quite a surprise in 1938 when they reappeared and proved us wrong.

So we've got our hypothesis, now let's look at some of the best known stories that the hypothesis is attempting to explain.

In the mid-1800's, so the story goes, hunters in the present day country of Georgia captured a wild woman they named Zana. She is variously described as tall, muscular, strong, hairy, and extremely resistant to cold. She was kept in a village and gradually became more domesticated, though she never learned to speak. Zana had a number of children, presumably as a result of her fondness for drinking great quantities of wine and falling into a deep sleep. She died about 1890 and was buried in the village. A number of Russian researchers have followed up on the story of Zana. Though no evidence survives of her existence, there were plenty of interviewees who knew Zana's children. She had four children who survived, all of whom grew up as relatively normal humans, though they were said to have dark skin and great strength. Only one burial site could be located, that of a son named Khwit who died in 1954. The skull is believed to be at the Moscow State University Institute of Anthropology. A National Geographic television program performed DNA testing on Khwit's skull, and found it to be conventional *Homo sapiens.*

A Shanghai newspaper reported in 1988 that hair samples collected from wildmen in the mountains of central China are definitely not human. China has a long association with wildmen; in fact, in 1976 the Chinese Academy of Sciences sent a team of 110 people to try and capture one that was said to have accosted six government officials. China also has its own story of a woman who was attacked by wildmen in 1940 and bore them a child, who was said to have dark skin and great strength. During World War II, Mongolian servicemen patrolling the Chinese border opened fire on what they supposed was a group of saboteurs, but found to their surprise that they had killed apelike creatures. Unfortunately, if this actually happened, nobody thought to photograph it or save any samples. Like Zana, the Chinese wildmen are generally said to be large, muscular and hairy, a partial match for what we know of Neanderthals. In fact it sounds more like a *Gigantopithecus.* If Zana was a *Gigantopithecus,* she could not have borne children from a human father; the genomes are simply too different. But what if she was a Neanderthal? It's not impossible. The Max Planck Institute is actually in the process of assembling a Neanderthal genome, from a fragment of femur found in Croatia in 1980. Until the genome is complete we won't know for certain whether it was possible for *Homo sapiens* and *Homo neanderthalensis* to interbreed. Most DNA testing done so far indicates that if humans and Neanderthals ever did interbreed, it was not at a significant level.

So now we've got a fair handle on the landscape of evidence in front of us, and now we can take a skeptical look at what we've got. Basically, nothing. We have some vaguely plausible hypotheses — yeah, I suppose it's possible that relic Neanderthals and *Gigantopithecus* or even some descendant of *Paranthropus* could survive in remote parts of Asia — but that's all we really have, a hypothesis. A hypothesis is a provisional explanation for these stories of wildmen in captivity and bearing children. There are stories of Almas from all over Asia from all time periods, just like we have reports of Bigfoot in the American Northwest. What's lacking is evidence that can be tested. The existence of Khwit's skull, and yet the lack of trumpeted headlines in the scientific journals, suggests that nothing remarkable has been learned from the skull, thus lending zero support to the Zana story.

Skepticism is not out to prove a negative, so I'm not going to say that the Almas does not exist. Science looks at the evidence, and makes a conclusion. With the Almas, we have an absence of evidence, which is not evidence of absence. By not expressing a belief in the Almas, critical thinkers are not being closed-minded. Indeed, we remain extremely open to any evidence that can be presented. DNA testing and genomes make this all much easier and more precise, so bring it.

40. THE ALIEN INVASION OF PHOENIX, ARIZONA

Tonight we're going to grab our shotguns, jump into our rambling pickup truck, and chase a massive triangular UFO as it courses silently across the American southwest, for we are (once again) on the trail of the infamous Phoenix Lights.

Perhaps it's the recent 10-year anniversary of the event, or perhaps it's the former Arizona governor's recent confession that he believes they were actually an alien spacecraft, but the Phoenix Lights have been back in the news again. It was the night of March 13, 1997 when a slanting line of bright lights appeared one-by-one in the sky beyond Phoenix, Arizona. Hundreds of photographs and videos were taken by observers throughout the region, making it among the most documented UFO sightings ever. The incident came as no surprise to anyone at nearby Luke Air Force Base (named for World War I ace Lt. Frank Luke), which operates the Barry M. Goldwater Range where a flight of four A-10 ground attack aircraft were jettisoning leftover illumination flares. The flares are typically dropped at lower altitudes, where they are not visible from Phoenix, due to the intervening Sierra Estrella mountain range.

The Phoenix Lights incident is a running joke in the Air Force and especially at the 104th Fighter Squadron of the Maryland Air National Guard, whose aircraft were involved. They don't have desert bombing ranges in Maryland, so the pilots go to places like Arizona for some of their training. The Air National Guard is the Air Force's reserve unit, similar to the Army Reserve.

But the rest of us regular people didn't know anything about this. We all just looked up into the sky, and saw something unlike anything we'd ever seen before. I remember watching it on the news with my wife. I remember my sense of amazement at witnessing something truly unexplainable: Could this actually be alien spacecraft?

Over the next couple of weeks, corroborating reports flooded in, of triangle-shaped craft from as far away as Henderson, Nevada cruising over the southwest, to Prescott, over Phoenix, and off toward Tucson. UFO's are reported nearly every day in most areas by someone, so it's to be expected that the normal background noise of typical reports would be given special attention during a large-scale episode like the Phoenix Lights. And, obviously, such a furor offers an easy opportunity for any clown to go on the news to say that a triangle-shaped craft passed over his house on its way to Phoenix. What *would* have been truly unusual and shocking is if there had been *no* other reports from nearby areas. Too bad none of *these* people owned cameras.

Lots of people in the Phoenix area *did* own cameras, and they all filmed exactly the same thing. Hundreds of photographs, hours of video, and all of it showing a line of lights in the sky above the city lights of Phoenix, looking toward the Sierra Estrella mountains and the Goldwater Range. Not a single photograph or frame of video showed anything else. This was the most documented UFO sighting in American history, and every last photograph showed exactly the same thing. Plenty of verbal reports told very different stories over the weeks following the incident, but every single photograph showed a simple line of lights beyond the Sierra Estrella.

As has been thoroughly documented, including by a Fox television special, the moment that each light disappeared on the evidential videotapes corresponded exactly with the horizon line of the Sierra Estrella mountains, proving that the lights were behind the mountains, and not over Phoenix.

Here's a story that's typical of the many found on the Internet, from Jan Markham of Gilbert Arizona:

> *My husband and I were out flying that night in the vicinity of the Stanfield VOR. We clearly saw the flares to our west, over the Goldwater range - a familiar sight to my husband. However, there was a second set of lights that night - the V-shaped formation that was initially shown on film by*

the local TV networks. That formation, whatever it was, flew directly over us at a much higher altitude than the flares. At the time, we thought it was some sort of military flight, but that never appears to have been acknowledged. I am sure someone knows the truth about those lights, but, please, don't insult our intelligence by telling us they were flares.

Let's spend a moment examining the flare said to be used in the incident. The A-10 drops two different kinds of flare: a countermeasure flare, used to confuse heat-seeking missiles; and an illumination flare, used to light up the ground at night either for the benefit of troops on the ground or to light up a target so it can be visually targeted for weapons release. The illumination flare is the one we're talking about. It's called the LUU-2 air-deployed high intensity illumination flare. It's made by defense contractor ATK Thiokol. The variant in use at the time of the Phoenix Lights incident was the LUU-2B/B. It weights 30 pounds and its canister is three feet long and 5 inches in diameter. Once it ejects its parachute and ignites, it puts out 1.8 million candela for 4 minutes, or 1.6 million candela for 5 minutes. It falls in its parachute at 8.3 feet per second. At 1000 feet above the ground, it lights up an area half a kilometer wide at 5 lux. The LUU-2's pyrotechnic candle burns magnesium, which produces an intense white light. Because it burns so hot, it also ends up burning the aluminum canister, which adds an orange hue to the light for most of the burn. About halfway through the burn, enough of the canister has been burned away that it actually lightens the load and it falls more and more slowly. Once it's almost completely out, an explosive bolt disconnects the parachute and the flare drops, burning out completely sometime hopefully before landing on someone's wood shingle roof.

The Barry M. Goldwater Range is a big place — over 4,000 square miles — and the Phoenix metropolitan area is even larger, about 14,000 square miles. The distance between the two is usually cited at 60 to 80 miles, but as we can see, that's going to depend on a lot. We do know a little about where the A-10's were flying inside the Goldwater Range. The guy who was in the lead A-10, Lt. Col. Ed Jones, says they were near Gila Bend when they ejected the leftover flares, and Gila Bend is just about exactly 50 miles from downtown Phoenix. Mesa and Scottsdale are farther away, so let's take a super rough stab at it, be conservative, and say that the average observer of the Phoenix Lights was 70 miles away from the A-10's. The brightness of the LUU-2 seen from 70 miles away is roughly equal to a star with an apparent magnitude of somewhere between -3.2 and -4.3, which is significantly brighter than any stars visible in the sky, but not as bright as the full moon. The magnitude scale was developed by the astronomer Hipparchus, where +1 represents the brightest star in the sky, and +6 represents the faintest. -3.2 is quite a bit brighter than the brightest star. The noonday sun has an apparent magnitude of -26.7. Thanks to the guys on the Bad Astronomy and JREF forums who helped me with these calculations.

Yet another wrench in the machinery is that all of the above is dramatically affected by atmospheric conditions. It wouldn't take much haze for absorption and scatter to obscure flares completely at that distance, and in the clear conditions predominant over Phoenix, lights are often distorted by an inversion layer, an effect that you can sometimes see when the landing lights of aircraft approaching an airport appear much bigger than they actually are. So we have a computation based on multiple unknown variables, any of which could wildly throw off our results. The one thing we can say with certainty is that the approximate brightness of the Phoenix Lights as seen in the photographs and videos does fall well within the wide range of brightness that's possible from LUU-2B/B flares at 70 miles.

Here's one final fly in the ointment. The photographic evidence itself is not necessarily a valid representation of how the lights would have looked to the naked eye: Still and video cameras are of varying quality and need specific settings to capture lights in the night sky. We have little or no information about the settings used in most of the available photographic and video evidence. Much has been made of a ham-handed spectral analysis of Phoenix Lights photographs and videos by prominent UFO advocate Jim Dilletoso, whose conclusions have been widely discredited since you can't even remotely do a spectral analysis of lights in a photograph and expect there to be any useful similarity to the spectrum of the actual light source, any more than you could expect a photograph of an orange to smell like an orange. Dilletoso found that, based on the colors in photographs, the Phoenix Lights could not have been from any known earthly source. Note that among Dilletoso's other claims to fame is having spent six weeks at an underground alien base in Dulce, New Mexico. Judge his credibility for yourself.

The UFO crowd and conspiracy theorists point out other problems with the flare explanation, most notably that a public relations officer at Luke Air Force Base contacted that night didn't happen to know that flares had been dropped, and so had no explanation for the lights. For this to be a real problem, you have to assume that everyone involved in training exercises immediately communicates every tactical detail of what they do, and their own personal estimation of its possible consequences, to the base PR officer. The officer also said that the Air Force had no operations over Phoenix that night, which was of course completely true. The A-10's were a great distance away and well inside their Military Operating Area airspace. This statement has been taken by the conspiracy theorists as evidence of a conspiracy, so discussing it is just beating a dead horse. The only other dissenting evidence put forward is the mass of eyewitness accounts following the triangle shaped craft on its journey across the southwest. Unfortunately all such stories are in direct contradiction with all photographic evidence. These witnesses had as much opportunity to document their sightings as did the people in Phoenix. The fact that they did not must be met, unfortunately, with a shrug. There are simply too many other reasons they might be saying what they're saying, and their reports are precisely contradicted by a mountain of hard evidence.

The Phoenix Lights were flares. Deal with it.

41. WHACKING, CRACKING, AND CHIROPRACTING

In this chapter we're going to lay down on the table, hold tight and grit our teeth, receive what a chiropractor once eloquently described to me as the "Whack & Crack," and have the flow of New Age energy improved through our bodies and spirits. Our next topic is chiropractic.

Like so many non-evidence based alternative medicine systems, chiropractic was established and defined by a non-scientist during a time when almost nothing useful or true was known about medicine. In this case, our inventor was Daniel D. Palmer, a practitioner of magnet healing, when medicine was in the Dark Ages of 1895. Palmer believed that his magnets could manipulate a type of immaterial spiritual essence which he believed exists in the body, and which he called "innate intelligence." Palmer reasoned that innate intelligence flows through the body through the nervous system, and that whenever an illness exists, it must be due to a nerve blockage preventing the flow of innate intelligence. It seemed reasonable to Palmer that straightening the spine through manual manipulation would relieve any crimps, thus curing virtually any disease and restoring health. Palmer called his new invention chiropractic, from the Greek for "done by hand."

Chiropractic's entire history has been quite stormy, and the early days were no exception. Palmer was soon arrested and convicted with practicing medicine without a license. His son, BJ Palmer, formed the first professional chiropractic association to cover legal expenses of the students he and his father trained.

Chiropractic is relatively unique among alternative medicine systems because, although it was originally developed based on the purely mythical and supernatural conjecture of innate intelligence, the profession as a whole has evolved and generally accepted most anatomical discoveries of modern medicine. Most (though certainly not all) modern chiropractors do accept many of the fundamentals of orthopedics and physical therapy. This has inevitably resulted in several different branches of chiropractic, with different sets of beliefs, and we'll talk more about those in a moment.

The cornerstone of chiropractic is something they call a subluxation. The first and most important thing to understand is that a chiropractic subluxation is a completely different phenomenon from an orthopedic subluxation, which is a real medical condition, and is unrelated. An orthopedic subluxation is a partial dislocation of a joint. They are significant physical displacements, and as such, they can and do appear on images such as X-rays, MRI's, and CAT scans. A chiropractic subluxation, on the other hand, is theoretic and is not visible on an imaging study or otherwise verifiable through conventional medicine. The chiropractic profession has repeatedly redefined a subluxation over the years, and the current definition is "a complex of functional *and/or* structural *and/or* pathological articular changes that compromise neural integrity and may influence organ system function and general health." As you can see, it's quite a vague definition and leaves plenty of room for individual interpretation. In practice, it usually refers to an alleged misalignment of adjacent vertebrae. According to the medical profession, such a misalignment would not have any of the detrimental effects on organs or general health claimed by chiropractors. Additionally, were there an actual nerve impingement in the spine, it would absolutely be visible on an imaging study and would absolutely not be treated through manipulation, which could easily result in irreparable injury. Therein lies the essential conflict between conventional medicine and chiropractic. Chiropractic treats imaginary conditions, that could not possibly cause the reported symptoms even if they did exist, using methods that would be highly detrimental on an actual impingement.

With such necessarily vague definitions, there are about as many different types of chiropractic as there are chiropractors; and indeed, most chiropractors do not belong to any sort of professional chiropractic association. However, most do fall into one of three main groups: Straights, Reforms, and Mixers.

Straights are those who stick firmly with Palmer's original concepts of innate intelligence, tend to reject modern medicine, and honestly believe that spinal manipulation can cure most any disease. Significantly, no evidence has ever shown that straight chiropractors have a lower incidence of any given disease, or of disease in general, which kind of makes you wonder. Reforms are the opposite. They accept that innate intelligence is not a real phenomenon and tend to restrict their treatment to types of manipulation that correspond with conventional physical therapy. Those few chiropractors who are also MD's are usually Reforms. The largest group of chiropractors are the Mixers, who, as their name suggests, attempt to marry some of Palmer's original ideas of subluxations with modern anatomical knowledge and treatments. Mixers often offer various other alternative medicine systems and often take a holistic approach to health. After many decades of controversy and licensing debates, there are now accredited colleges through which chiropractors can become licensed to practice. A Doctor of Chiropractic is not a medical doctor, and is not licensed to prescribe drugs or to perform surgery in the United States. They can be listed as primary care providers, which seems surprising given they are not trained or allowed to do something as simple as prescribe an antibiotic or set a broken bone.

I have some volleyball friends who see chiropractors regularly, and swear by them. Like some other sports, volleyball is one that keeps its elder players fairly constantly in the offices of orthopedic surgeons and physical therapists. Athletic massage and physical therapy are often essential parts of injury recovery, but if improperly performed, they absolutely have potential to cause more damage and make a bad situation worse. That's why we have certification boards for massage therapists and Doctors of Physical Therapy — top physical therapists should have a DPT after their name on the door. Physical therapists who are not doctors still must have taken an accredited four-to-six-year college program and must pass a national physical therapy examination and an examination on the laws and regulations governing the practice of physical therapy. Physical therapy assistants must take an accredited two-year college program and must pass the national physical therapist assistant examination, and they may only work under the supervision of a licensed physical therapist. A physical therapy aide is not licensed and is not required to meet any education requirements and has no formal training. However, they are required to work only under the direct physical supervision of a licensed physical therapist. When my volleyball friends report back about what their chiropractor did for them today, guess what? It's often exactly the same treatment I've received from my DPT. Some of these chiropractors are doing conventional physical therapy but without having taken the training and passed the tests, and they're getting away with it because they're calling it chiropractic. Not only is that untrue, it's illegal, unless that chiropractor also happens to be a licensed physical therapist. If you have a painful sports injury, you should be going to an orthopedist anyway, who is licensed to provide medical care and can do things like order an MRI to properly assess an injury.

Many chiropractors are rational people and are knowledgeable about sports medicine or back pain, and do provide good physical therapy. The best will often be openly critical of Straight chiropractors and advise you to avoid any practitioner who follows the subluxation philosophy. This is good, but it's not as good as receiving the same advice from someone who went to medical school and whose practice is built on medical science. My question to these Reform chiropractors is: If you are so critical of the chiropractic arts, then why are you a chiropractor yourself? If you want to be a doctor and help people, fine; go to medical school, and become a doctor. Yes, it's easier, cheaper, and faster to go to chiropractor school, and there isn't so much pesky "anatomy" to learn, but if you believe medical services should be based on medical science, then you should go all the way. I'm tired of hearing chiropractors be critical of chiropractic. It's the pinnacle of hypocrisy.

There's one criticism of chiropractic that I'm not going to urge, and that's the fact that these spinal manipulations can be extremely dangerous and can cause spinal injuries that have resulted in paralysis and deaths. The most common injury is a stroke following neck manipulation. The reason I'm not going to urge this criticism is that mistakes can be made in every type of medicine, whether it's alternative or conventional. A pharmacist friend of a friend once made a mistake, filling a prescription with the wrong medication, and a child died as a result. During the ensuing lawsuit, the pharmacist took her own life. No type of treatment is free of the risk of accidental error. Fortunately, they're extremely rare.

If you have some medical condition that you've been treating with chiropractic, consider going to a medical doctor for a proper diagnosis. If an athletic massage or physical therapy are prescribed by your doctor, go to a proper physical therapist or licensed massage therapist, who are able to give you better treatment, legally and with the proper training under their belt, and who understand the medical basis for their treatment. You can only do better than with a chiropractor whose training is founded upon Palmer's 1895 conjecture of innate intelligence.

42. A Mormon History of the Americas

Join us now as we slip into our magic underwear and enter a mysterious building that no outsider has ever visited — a Mormon temple — for we're now studying that most curious of history texts, the Book of Mormon.

The Mormons, formally and properly known as the Church of Jesus Christ of Latter-Day Saints, is the same as any other more conventional Christian church, but with the additional element of a belief that after the Resurrection, Jesus also appeared to peoples in the Americas. The story goes that a man in Jerusalem named Lehi built a boat for his family and sailed across the Atlantic to the American continent in about 600 B.C., and they became the forebears of the American Indian people of North and South America. The history of the continent was kept on gold tablets, passed from generation to generation, compiled by a warrior named Mormon and finally buried in upstate New York by his son Moroni. A golden statue of Moroni, now an angel and holding a long trumpet, stands atop most Mormon temples — all unofficially pointing toward Jackson County, Missouri, which Mormons believe is the geographic center of the continent, and where they believe Jesus will make his Second Coming. That's a free tip for you property investors.

Now the early days of the Mormon church were violent. It all began around 1827 when a young man of 22 named Joseph Smith revealed that Moroni had been appearing to him in dreams for some time, and had guided him to the location of the buried gold plates. With divine guidance, he translated the plates from the "modified Egyptian" in which they were written, published the text as the Book of Mormon, and begun to acquire followers. This was a tall order in those days of staunch Protestant Christianity, and the early days of the church were bloody indeed. Whole wars were fought in counties throughout Illinois and Missouri, and it was some decades before the Mormons decided enough was enough, and were led by Brigham Young to the safe haven of Utah, where they founded their kingdom called Salt Lake City, and got to work building some of our finest ski resorts.

Once we get past their early years, when murders and even massacres were committed by both sides, what you'll find to be generally true of Mormons today is that they are among the most upstanding of citizens. They generally don't drink or smoke, crime is almost unheard of, they have great family values, and if you believe Playboy magazine, BYU women are among the hottest in the nation. It's true that toward the end of his life, Howard Hughes kept his inner circle composed largely of Mormons, not because they never drank as some stories say, but because he felt they were the only people who were truly trustworthy. You could do a lot worse than Mormons if you want good next-door neighbors. They even use pooper scoopers.

So what is there about the Mormons to be skeptical of? Well, it's not the polygamy, which the church gave up as a condition of statehood in 1896. Certainly nobody who believes in the Bible should have a problem with polygamy, and most of the rest of us couldn't care less how many wives other people want to have. It's not even the whole thing with the gold plates, evidenced only by a sworn testimonial from Joseph Smith's closest confidants who claimed, as Mark Twain noted, to have "hefted" them. It's not even that Joseph Smith couldn't possibly have written that much detailed and well-constructed stuff all by himself: Whether he did it himself or was assisted by his team of ghost writers doesn't prove or disprove anything about the accuracy of its contents.

The part of Mormonism to be skeptical about is the demonstrably untrue ancient history.

People who believe in Bible stories are on thin enough ice as it is, but at least a lot of them have enough sense to say that the stories are allegorical and not meant to be taken literally. Mormonism, on the other hand, claims that the history in the Book of Mormon is the correct history of the peoples of the American continents, no allegory involved. Yet, every falsifiable detail of the Mormon account has been easily shown to be completely untrue.

For one thing, the genetic evidence shows that native populations in the Americas came from Asia via the land bridge at the Bering Sea, not from Europe. American native populations fall into one of four haplogroups. Haplogroups are the main branches of the human genealogical tree, defined by markers on the Y chromosome and Mitochondrial DNA, and corresponding to early human migrations to the various continents. The consensus of opinion among biological anthropologists is that all four American haplogroups bear markers that tie them to Asia. There is very little dissent from this consensus, and what little there is comes mainly from fringe religious groups. Dr. Michael F. Whiting, a biologist with Brigham Young University's Maxwell Institute for Religious Scholarship, responds to the majority opinion thusly:

The first point that should be clarified is that those persons who state that DNA evidence falsifies the authenticity of the Book of Mormon are not themselves performing genetic research to test this claim. This conclusion is not coming from the scientists studying human population genetics. It is not the result of a formal scientific investigation specifically designed to test the authenticity of the Book of Mormon by means of genetic evidence, nor has it been published in any reputable scientific journal open to scientific peer review. Rather, it has come from outside persons who have interpreted the conclusions of an array of population genetic studies and forced the applicability of these results onto the Book of Mormon. The studies cited by these critics were never formulated by their original authors as a specific test of the veracity of the Book of Mormon. To my knowledge there is no reputable researcher who is specifically attempting to test the authenticity of the Book of Mormon with DNA evidence.

This is probably true, and the reason is that the Book of Mormon is not a scientific theory. If it was, research teams would be trying to test it and falsify it, to verify its validity.

Since it's a religious myth, there are about as many legitimately funded biologists studying it as there are zoologists trying to determine whether serpents can talk.

Evidence against the Book of Mormon is not just genetic. The Book of Mormon is full of references to technologies and species that are known to have not existed in pre-Columbian America. Michael Coe, an archaeologist at Yale University, said:

> *There is an inherent improbability in specific items that are mentioned in the Book of Mormon as having been brought to the New World by...Nephites. Among these are the horse, the chariot, wheat, barley, and [true] metallurgy. The picture of this hemisphere...presented in the book has little to do with the early Indian cultures as we know them.*

Mormon scholars do have answers to some of these questions. For example, they propose that meteoric nickel-iron alloy could have been mistaken for steel. FairLDS.org defends the Book of Mormon against the contradicting scientific evidence, in a series of lengthy essays full of scientific language, yet often citing the Bible as the authority for its assumptions. Well, it's all well and good to hypothesize all day long, but the only thing we can know for sure is what we find in the physical evidence. And all the evidence shows that many technologies and species described in the Book of Mormon were introduced to the continent in modern times, and that the native Americans all descend from Asian migrations many thousands of years before the Book of Mormon stories were said to take place.

There are no better next door neighbors than Mormons. No better examples of family values and clean, healthy living. But, you can be all of those things and have all of those things — including being a good Christian, if that's what you want — without insisting on the literal truth of a nineteenth century book that is not only improbable, but is exhaustively evidenced to be false.

43. THE TRUTH ABOUT REMOTE VIEWING

Next let's try sitting in a quiet room to draw sketchy pictures of — well, of anything, really — and claim psychic powers, for we're demonstrating the amazing psychic ability known as "remote viewing."

Remote viewing was made popular beginning in the 1970's, when some in the US intelligence community grew concerned that the Soviets had better psychics than we did. $20 million was appropriated to test the skills of a group of psychics called remote viewers. Supposedly, you could ask them a question about some place, and they'd use psychic abilities to draw you a picture of whatever's going on there, and it was hoped that this would lead to useful intelligence. Project Stargate, and a few others like it, was canceled by the 1990's, due to a lack of reliable results. Proponents of Project Stargate say that the US government's investment in the project proves that it had merit. Critics point out that the funding was stopped, and say that if merit had been found, funding would have at least been continued, if not dramatically increased. We can be reasonably assured that the project did not move underground with renewed funding, since the participants have all long since gone public with full disclosure of what happened. Since none of them have turned up mysteriously disappeared, we can safely assume that the government is not too concerned about this supposedly "classified" information.

The most famous remote viewer to emerge from these projects is a man named Joseph McMoneagle. Today he offers his remote viewing services on a consulting basis, and in 1994 he went on the television show *Put to the Test* to show just what he could do. There is a clip from the show on the Skeptoid.com web site, and if you want, put this book down now, go and watch it, form your own opinion, and then come back to read my comments. What you'll find is that the show's unabashed endorsement of his abilities contributes largely to the perception of his success, but if you really listen to the statements he makes, and look at the drawings he produces, you'll find little similarity to what he was supposed to identify. They took him to Houston, Texas and sent a target person to one of four chosen locations. McMoneagle's task was to draw what she saw, thus determining where she was. They edited the 15 minute session down to just a couple of minutes for the show, so you've got to figure that they probably left in only the most significant hits and edited out all of the misses.

The four locations were a life size treehouse in a giant tree, a tall metal waterslide at an amusement park, a dock along the river, and the Water Wall, a huge cement fountain structure. Here is what McMoneagle said:

❖ There's a river or something riverlike nearby, with manmade improvements. Houston is a famous river town, so this was a pretty good bet. It applies equally well to the waterslide and to the dock.

❖ There are perpendicular lines. I challenge anyone to find any location anywhere without perpendicular lines.

❖ She's standing on an incline. She was not standing on an incline, and there were no apparent inclines at any of the four locations. Remember, they edited it down to just the most impressive two minutes.

❖ She's looking up at it. This would apply best to the treehouse, the waterslide, or the Water Wall. There was really nothing to look up at at the dock – there was a traffic bridge some distance down the river, but up would not be the direction to look to see it.

❖ There's a pedestrian bridge nearby. Sounds like a close match for the treehouse or the walkways on the waterslide.

❖ There is a lot of metallic noise. Probably the big metal waterslide structure is the best match for this.

❖ There's something big and tall nearby that's not a building. This applies equally well to all four locations.

❖ There's a platform with a black stripe. Not a clear match for any of the locations.

That's it - those were the only statements of Joe's that they broadcast. Strangely, at no point did they ask McMoneagle to identify the location; they did not even ask him to choose from the four possibilities. Instead, they simply took him to the actual destination where the target person was, which turned out to be the dock, and then set about finding matches to Joe's statements. Suddenly, nearly all of Joe's statements made perfect sense! Certainly there's a river nearby. There was a traffic bridge in the distance: traffic, pedestrians, near, far, no big difference. Metallic noise and something big: there was a ship at the dock, but if you ask me what kind of noise a ship makes, metallic is not the word I'd use. And that platform with a black stripe? Could be a ship.

I argue that the target person could have been at any one of the four locations, and Joe's psychic predictions would have seemed equally impressive. Joe made numerous sketches, but the only two that they showed were a sketch of a squiggly river (the river at the dock is between straight cement seawalls) and a vague triangular shape, which they interpreted as similar to a crane on a barge when seen from a certain angle.

Bottom line: The only thing I found impressive about McMoneagle's demonstration was their editing and narration job to make it look like the most amazing and miraculous psychic feat in history. Maybe he failed this time because he was not in complete control of the test conditions, as he was in Project Stargate. Maybe the rest of time, McMoneagle is able to display spectacular unambiguous results. McMoneagle claims a decent hit rate, but not perfect. If I were a professional remote viewer, I too would claim a less-than-perfect success rate: High enough to sound impressive; but low enough to allow for potential failures in cases where protocols were imposed that I couldn't control.

I'm not a magician myself — it's really sad to even watch me try to shuffle a deck of cards — but I do know how a lot of the tricks are done. And I can assure you (more importantly, any professional magician can assure you) that the abilities claimed by remote viewers are well within the magician's bag of parlor tricks. This doesn't *prove* that remote viewers are just putting us on with simple tricks, but their claims and their results are consistent with that. Which of these two possible explanations is most likely true: That remote viewers are using well-proven techniques demonstrated by professional and amateur magicians every day; or that they are accomplishing a feat of true paranormal abilities, which has never been demonstrated under controlled conditions, cannot be duplicated by anyone else, and has no proposed mechanism by which it might be possible?

Now I'll be the devil's advocate, and give the reply that most believers in remote viewing are probably thinking right now: That my characterization is untrue, and that these feats of knowing the unknowable are performed under controlled conditions, and that magicians cannot duplicate these feats. I'll answer that now, and while I do, keep one thing in mind: that the "controlled conditions" under which Joe McMoneagle performed at Stargate were, according to him, defined and set up by Joe McMoneagle himself — thus putting the fox in charge of the chickens.

In 1979, Washington University in Missouri received a $500,000 grant from James McDonnell, of McDonnell-Douglas, to investigate psychic abilities. Noted professional magician James Randi secretly recruited two teenagers, Steve Shaw and Mike Edwards, and gave them a basic training in stage magic and the art of deception. To flatten the playing field, Randi also contacted the McDonnell researchers and suggested a set of protocols that would detect any such trickery as that with which he instructed his so-called Alpha Kids. He also suggested that they have an experienced magician present during their experiments to look for such techniques. Neither suggestion was followed. As a result, out of 300 applicants claiming to have psychic abilities, only Shaw and Edwards passed the preliminary examinations and were accepted into the program. For the next four years, Shaw and Edwards consistently amazed the researchers, and the parapsychology community at large, with their psychic abilities. Like McMoneagle, Shaw and Edwards were often allowed some amount of control over the conditions. Randi tried to confess the hoax by performing all the same tricks and explaining exactly how Shaw and Edwards were doing it, but the researchers didn't believe him. Randi finally laid it all out in Discover magazine, the research came to a stop, and there were widespread shockwaves throughout the parapsychology community.

After the conclusion of Project Alpha, Randi said:

> *If Project Alpha resulted in parapsychologists awakening to the fact that they are able to be deceived, either by subjects or themselves, as a result of their convictions and their lack of expertise in the arts of deception, then it has served its purpose.*

The lack of expertise in the arts of deception. Unfortunately, nearly all of us outside the world of professional magic lack such expertise. The inevitable conclusion to be drawn from Project Alpha is that magicians, even relative novices like Shaw and Edwards, can fool very serious researchers under controlled conditions, even when those scientists are serious about finding flaws in the methodology and looking for hoaxes, and even after having been briefed by Randi himself on what to look for. It is not hard to reach the corollary conclusion: That non-investigative, non-scientific, non-critical minds, like Joe McMoneagle's audiences and the people he worked with in the CIA, could also be duped by similar skills, and be firmly convinced of their reality. You want remote viewing? Steve Shaw, who now performs under the stage name Banachek, can read the ID numbers off a card in your pocket, and he can do it on stage every time, without any mistakes, without any outside assistance, no cameras, microphones, or other trickery involved.

When you see something that seems impossible, approach it skeptically. Before you accept that it's something outside of our world, first check to be certain that it's not already inside our world. The tricks used by remote viewers and the magicians who emulate them are definitely inside our natural, fascinating, amazing world.

44. The Importance of Teaching Critical Thinking

Now let's try something a little bit different. Rather than talk about any one specific phenomenon, I want to talk in general terms about the importance of teaching critical thinking to young people, and how and why it can and should be done better. A skeptical approach to life leads to advances in all areas of the human condition; while a willingness to accept that which does not fit into the laws of our world represents a departure from the search for knowledge.

We had a critical thinking class at my high school as an elective, and I think it was generally considered to be the most boring and useless class you could take. If memory serves, the bulk of the class involved reading and studying Plato's Socratic dialogs. If you read them as a teenager, you may recall your reaction was to find them pretty darn dry. They were dialogs between Socrates and other people about such riveting subjects as ancient politics, philosophy, and even mathematics. I don't mean to criticize Socrates; it's just that studying the man and his 2,400-year-old writings is about the least interesting and relevant way for a modern young person to get excited about what Socrates was communicating. Nobody I knew who walked out of that class ever remembered a single concept, or applied it to their life. You can disagree with me and say that you find the Socratic dialogs to be brilliant and fascinating. My point is that the average teenager does not.

But the concepts Socrates introduced, such as the Socratic questions, are brilliant and fascinating when we apply them to things that interest us. More significantly, they become relevant. Take a few Socratic questions:

❖ What is the source of your information?
❖ What assumptions are you making?

❖ Is a different conclusion more consistent with the data?

❖ What is an alternate explanation for this phenomenon?

What if we encouraged young people to ask these questions not of early Greek politics, but of the issues they're hit in the face with every day? Global warming. Television psychics. Alternative medicine. New Age religions. Popular assumptions about alternative fuels. Alternative foods. Alleged correlations between Xbox violence and actual violence. Magnet therapy. Isn't it more useful to encourage better ways to think about the subjects that people are already thinking about?

I have a favorite example of an older, less interesting critical thought exercise that was made more relevant and interesting. The philosopher Bertrand Russell, in criticizing major religions, conceived of what became known as Russell's Teapot, a small china teapot allegedly orbiting the sun. Since nobody could disprove its existence, Russell argued that the only reason its existence shouldn't be taken for granted is that there are no ancient texts written about it. He applied Socratic reasoning to point out that ancient texts do not constitute proof of an unprovable concept. Russell's Teapot was freshened when a grad student named Bobby Henderson wrote to the Kansas Board of Education in 2005, which had just mandated that the Christian story of creation should be taught instead of science. He insisted that by the same logic, his non-disprovable Flying Spaghetti Monster deity's myth of creation should be taught with equal time. Fans appreciative of Henderson's logic have since formed the parody religion known as Pastafarianism. The Flying Spaghetti Monster is goofy and glib, but it is a valid example of using critical thinking to analyze the value of a real phenomenon that we face today.

Teachers, what would your students come up with if you asked them to apply similar reasoning and invent an alternative to television psychics, founded upon the same assumptions that Sylvia Browne asks us to make?

Finding fault with television psychics or the Kansas Board of Education is not, by itself, a positive contribution. Skepticism should not be merely a negative influence. Skepticism is not about debunking, disproving, or ruining anyone's faith. Skepticism is about applying the scientific method to arrive at a conclusion that is evidenced to be beneficial, like curing cancer. If, during this process, it first becomes necessary to debunk an unsupported alternative that's in the way, such as treating cancer with magnets, then that debunking serves as a stepping stone to the final solution. Debunking should never be an end in itself, because that alone creates nothing useful. As scientists, we are interested in learning, and often that involves replacing an older hypothesis that's found to be wrong.

Some people criticize science by pointing out that it does not know everything and doesn't have all the answers. Case in point: the popular movie *What the Bleep Do We Know.* Obviously, this criticism is true. Science is *all about* the fact that we don't know everything. *Science is the learning process.* There are ideologies that do offer all the answers, often divine in nature or based on ancient philosophies. When you have all the answers, there is no longer any need to learn, and thus no use for science. If we want to improve the world, improve the human condition, improve technology; learning, and thus science, is the essential way forward. Ideologies that offer all the answers are the essential route to developmental stagnation. When your hear someone criticize science because it doesn't have all the answers, don't argue with them; instead point out that that's the *strength* of science. We couldn't be learning more every day if we presumed to already know everything.

Some people criticize skepticism because it doesn't leave well enough alone. Many paranormal beliefs and alternative systems, even though they may lack hard scientific evidence, bring comfort to those who practice them and are a positive force in many peoples' lives. There is value and enlightenment to be found in life that isn't necessarily found in a science book. It is often argued that skepticism is not merely unimportant, it can even be harmful. Young people should not complacently accept this short-sighted argument. First of all, happiness and enlightenment are all around us in our world; they are not found only within a given pseudoscience. But moreover, once we begin investing our faith in unsubstantiated or supernatural phenomena, we are contributing to the redirection of attention, influence, and funding *away* from technologies and concepts that *have* been evidenced to be beneficial to humanity and to our world. As Yoda says: "If once you start down the dark path, forever will it dominate your destiny; consume you, it will." The choice between pseudoscience and science is the choice between stagnation and progress: Progress toward long life, health, happiness, a cleaner planet, bountiful food, knowledge, and peace.

45. Support Your Local Reptoid

Collect your children and run for cover. We're going to look at the terrifying tale that says a race of tall reptilian beings lives among us, and even runs our government.

The concept of reptilian beings on Earth is a surprisingly widespread conspiracy theory, in which the US government and major public companies are complicit in a vast worldwide network of underground bases housing a large population of humanoid reptilian creatures called Reptoids. They speak English and are involved in every major government and corporate decision. They are variously said to either disguise themselves or actually shape-shift into humans, where they have public lives in positions of national importance. Some say the Reptoids are of extraterrestrial origin, and some say they are native to Earth, having developed intelligence before the primates, and have been secretly running things all along.

I first heard of reptilians when planning a trip to Mt. Shasta as a youth. Shasta is one of our fourteeners here in California. As I discovered, it's also something of a sacred hotbed for a whole range of New Age traditions. It not only has a lot of Native American spiritual history, it also figures prominently for any number of modern pagan religions. Shasta is said to be full of secret caverns, jewel encrusted tunnels, and whole subterranean civilizations peopled with all sorts of exotic races. Most notably, it's the home of the Lemurians, an ancient race whose original continent called Mu sank and now make their home inside the mountain, in the great five-level city of Telos. Lemurians, who are tall, white-cloaked beings speaking English but with a British accent, employ invisible four-foot-tall beings called Guardians to protect their city. Bigfoots are also said to populate Shasta. Among all this exotic company, Reptoids would hardly be noticed. The story goes that Reptoids use Mt. Shasta as one of the numerous entrances to their huge underground network of bases.

Reptoids are said to serve at least one very useful purpose: They are sworn enemies of the gray aliens, and may well serve to be humanity's last line of defense against this threat. Among the gray aliens' holdings provided them by the US government is a large underground base at Dulce, New Mexico. Some 18,000 grays are said to reside on level 5 of the base, and they perform terrible genetic experiments on humans on levels 6 and 7. Reptilian beings have been caught trying to acquire information about the Dulce base.

The most outspoken proponent of the conspiracy theory that reptilian beings in disguise are actually running our planet is David Ickes, whose book *The Biggest Secret* reveals information like this:

> *Then there are the experiences of Cathy O'Brien, the mind controlled slave of the United States government for more than 25 years... She was sexually abused as a child and as an adult by a stream of famous people named in her book. Among them were the US Presidents, Gerald Ford, Bill Clinton and, most appallingly, George Bush, a major player in the Brotherhood, as my books and others have long exposed. It was Bush, a pedophile and serial killer, who regularly abused and raped Cathy's daughter, Kelly O'Brien, as a toddler before her mother's courageous exposure of these staggering events forced the authorities to remove Kelly from the mind control programme known as Project Monarch.*

This is a fair sample of most of Ickes' evidence that reptilian beings have taken over our government. Virtually any statement that Ickes makes is easily falsified by minimal research if not simple common sense, but since his is a conspiracy theory, any evidence against it is simply regarded as evidence proving the conspiracy. Don't laugh: Ickes sells a lot of these books. A lot of people believe this stuff.

Where did all of these stories come from? The earliest reference I've come across is from a *Los Angeles Times* news story from January 29, 1934, which is available from the *Los Angeles Times* archives. Geophysical mining engineer G. Warren

Shufelt had been using "radio x-ray" and had discovered subterranean labyrinths beneath the city of Los Angeles, including pockets of pure gold, and taken x-ray pictures of many of the chambers. Somehow Shufelt met with a man named L. Macklin, said to go by the Hopi Indian name of Little Chief Greenleaf. Macklin told Shufelt of a Hopi legend of Lizard People, an advanced race, who built the city beneath Los Angeles to escape surface catastrophes some 5000 years ago. Their history was kept on gold tablets. It sounded like Shufelt had struck paydirt — almost. He still had to dig it up. Shufelt's crew dug a shaft 250 feet deep, well below the water table, which of course promptly filled with water, and that's where the story came to an end.

So I began looking into the various elements from the *LA Times* story. First on the list was Shufelt's "radio x-ray" device. *Times* reporter Jean Bosquet described it:

> *Shufelt's radio device consists chiefly of a cylindrical glass case inside which a plummet attached to a copper wire held by the engineer sways continually, pointing, he asserts, toward minerals or tunnels below the surface of the ground, and then revolves when over the mineral or swings in prolongation of the tunnel when above the excavation.*

So, it turns out, Shufelt's device has little to do with either radio or x-rays and more to do with a common dowsing pendulum. This was all he had to guide his elaborate drawing of the catacomb layout.

> *Shufelt stated he has taken "x-ray pictures" of thirty seven such tablets, three of which have their southwest corners cut off. "My radio x-ray pictures of tunnels and rooms, which are subsurface voids, and of gold pictures with perfect corners, sides and ends, are scientific proof of their existence," Shufelt said.*

Shufelt's dowsing results notwithstanding, parts of the story seem unlikely. Gold, and metallurgy in general, was unknown among the Hopi until the mid 1700's. So was chemistry, but Macklin said that the Lizard People "perfected a chemical solution by which they bored underground without removing earth and rock."

I did make a pretty thorough effort to track down any such Hopi legend, but came up empty handed, not counting numerous modern references to Mt. Shasta and the Los Angeles catacomb story. I did find a "Lizard clan" referenced in several Hopi stories, but always among other clans (the Spider clan, the Bear clan), and never any references to underground cities, golden tablets, or any other elements from Shufelt's story. Obviously, my failure to find any evidence of such a legend doesn't prove anything: Native American legends were traditionally passed by word of mouth and never were written down, the only exceptions being those that made it into modern storybook collections. I was also unable to find a man named either L. Macklin or Little Chief Greenleaf in the public birth and death certificate databases for the Hopi Reservation in the Navajo Nation Court, but again, all this proves is that I didn't find it.

If Shufelt's dowsing misadventures truly were the genesis of modern Reptoid legends, there is an ironic aspect. Macklin never said that there was anything reptilian about the Lizard clan, they were simply one subculture of the Hopi, though just as human as anyone else. According to the story Macklin told Shufelt:

> The Lizard People, the legend has it, regarded the lizard as the symbol of long life. Their city is laid out like a lizard, according to the legend, its tail to the southwest ... its head to the northeast.

Most likely, this tall tale from the early days of Los Angeles was little more than an effort by Shufelt to interest investors in his treasure hunt, in which he no doubt believed wholeheartedly. As for Macklin? Who knows, Shufelt could have made him up, or he could have been a real guy, possibly even a real Hopi, and may have even told a genuine — if undocumented — Hopi legend. What Shufelt didn't know was that his little gem in the *Los Angeles Times* was the kickoff for a whole generation of one of our most bizarre (and entertaining) urban legends.

46. Free Range Chicken and Farm Raised Fish

Let us now sit down for a meal and compare free range chicken to regular chicken, and farm raised fish to regular fish caught from the ocean. Is either morally better? Is either healthier? The truth may surprise you. Few people have actually looked into the facts personally.

Let's look at the legal definition of "free range" as far as chickens are concerned. According to the US Department of Agriculture, free range chickens are simply those which have access to the outdoors. There is no clear definition of outdoors, however. Free range chickens are still fenced in and typically have a roof over their head as well, but conditions are as varied as there are numbers of farms. No doubt there are some smaller producers who raise chickens in the way that animal activists imagine free range chickens living, with wide open spaces and happiness and joy, but they are in the minority, since that's such an inefficient use of space. The vast majority of chickens sold as free range are simply given some access to outdoor space in approximately the same proportion that their higher market price justifies any reduced farming efficiency. Often it means little more than a window, and that's perfectly legal. Note that free range chickens have nothing to do with organic standards. Free range chickens can be organic or non-organic. That all depends on the food they're given and whether or not they receive antibacterial treatment, plus a few other details.

Proponents of free range chickens have been known to criticize buyers of regular chickens for the immorality of raising chickens in pens. I encourage those people to actually look into the facts of what free range means. It does not mean what most people think it does.

There are two main issues that free range chicken proponents wish to address: The well being of the chickens, and the healthfulness of their meat.

Let's talk about the well-being of the chickens. Does the freedom to walk or look outdoors give them a happier life? We raised chickens when I was a kid, and one thing we always thought was fun was to catch a chicken, lay his head on the ground, then put your finger at his nose and draw a line in the dirt away from him. We called this hypnotizing the chickens. Once you got the hang of it, you could let go of the chicken, and he'd lay there frozen for minutes, sometimes even longer. A lesson that I learned thoroughly was that chickens are not the most intelligent animals on the planet. Sometimes they'd be in the shed, sometimes they'd be out of the shed, always they'd be walking around clucking and pecking at stuff on the ground trying to eat. That's really all they did. I personally spent enough time around chickens to feel assured that a chicken's life is no richer when he's outdoors than when he's indoors. My personal assessment of free range chicken proponents is that they either did not spend as much time around chickens as I did, and believe them to be somehow enriched by that great outdoors feeling; or that they have some other chicken-communing experience outside of my own.

And so on to the second point: Is the meat of free range chickens healthier to eat? Remember, we're talking about free range chickens, not organic chickens, so this has nothing to do with what the chickens eat or what other treatment they receive. Some say that free range chickens get better exercise, so their meat is leaner, but this is simply untrue in most cases. Chickens sold as free range rarely have more space than regular chickens, the only difference is that some of that space is outdoors. This question really comes down to Salmonella. Some proponents say that the outdoor environment is free of the concentrated filth found indoors, and thus there is less bacteria; opponents say that the indoor pens are frequently sterilized for just this reason and are thus far cleaner than unsterilized outdoor chicken pens. There are probably cases where each of these is true to some degree. According to the research published on PubMed — the online medical research database published by the National Institutes of Health — there is no significant difference in the number of Salmonella found in conventional, free range, or organic chickens. You are just as likely to have Salmonella in your chicken no matter which you buy. So cook your chicken all the way through no matter what.

In the United States, free range chicken *eggs* are not regulated; they do not need to come from free range chickens, it's an unregulated marketing label only. There are no requirements which must be met by producers who sell eggs as free range, and so paying these higher prices is just throwing money away. Know what you're paying for. Do your own research.

Personally, I find myself without any reason to pay the higher prices for chicken marketed as free range. I doubt that the living conditions are actually significantly better, I doubt that chickens have the capacity to appreciate any difference there might be, and I am satisfied that free range chicken contains no less Salmonella.

What about hatchery raised fish, also known as aquaculture? This is a more complicated issue, because fish are difficult and expensive to get out of the ocean, and there are certainly cases where overfishing threatens wild populations. In this sense, fish farms make all the sense in the world: the native populations are not affected, and the fish can be harvested far more cheaply, efficiently, and safely.

This is a different question from the one about raising fish in hatcheries in order to help repopulate depleted stocks, which is a particularly thorny environmental issue. Conceptually it's a good thing to do, but in practice it creates highly complex problems. Releasing large numbers of fish into an area with multiple threatened species will help the released species, but often to the detriment of the other species. That's not to say it shouldn't be done, it just has to be done with great care by knowledgeable experts.

Fish farming is considered a good thing by such a large consensus that you have to dig pretty deep to find criticism of it: You have to dig all the way down to our favorite anti-human fire-bombing eco-terrorists at PETA, just the people you want in charge of your unbiased science information. They've made a web site called FishingHurts.org where they refer to the water in fish farms as "fecal stew" and actually presume to authoritatively discuss the psychological damage suffered by hatchery fish. They describe fish as "intelligent and interesting individuals". They also argue, strangely, that eating fish is toxic. That's news to me; I eat as much fish as anyone and I appear to be alive. Everyone is well within their rights to believe PETA's charges. If you do, you're probably not going to eat fish from any source, and so the question of whether it's better to eat farm-raised fish or free-swimming fish is not at issue.

And even back on Earth, when you're raising fish to eat, fish farming is not all upside. There are still problems. Since the populations are smaller, they are subject to inbreeding depression, so it's necessary to continuously introduce new genes into the environment. It can't be a totally closed system, but this very small draw on the wild population is far, far better than a direct draw on the wild population for all harvesting.

With a landlocked fish farm, growers have direct control over the water quality and content. No doubt there are fish farms where the water quality is deleterious to the fish, but since this hurts the farmers more than anyone else, they're in the minority. There are also bodies of natural water containing pollutants that fish in farms are not exposed to, which is to the farmers' advantage. Italy is really big on aquaculture, and those interested in the subject are encouraged to read up on their tests, which are numerous. Generally they find safe levels of many pathogens in both land-based and offshore fish farms, but no Salmonella in either. PCB's are found in slightly higher concentrations in the offshore fish farms, but still at safe levels. Again, you can find this information online at PubMed, which is a great bookmark for anyone interested in health sciences. Just don't tell anyone at PETA; we wouldn't want to pollute their minds with any of this immoral "research".

Overfishing in the oceans is a real challenge, but the severity of the problem and how recoverable it is depends entirely upon who you ask. I'm not even going to go there, that's another subject for another time. The bottom line is that fish farms are generally a good thing: They protect wild populations while still providing the fish we need. As for those free range chickens? If you're really concerned about the welfare of the chickens, don't eat them. There is little reason to conclude that chickens sold as free range under our current USDA standards live more fulfilling lives than their indoor counterparts. If you think they should, then you should probably direct your efforts toward changing the regulatory system, rather than criticizing people who eat regular chicken.

47. THE BIBLE CODE: ENIGMAS FOR DUMMIES

In this chapter we're going to load some cheap software onto our laptop and decode that silliest of modern pop phenomena, the Bible Code.

In 1994, an American journalist named Michael Drosnin visited Israel and told a poet friend, Chaim Guri, that he had a letter for prime minister Yitzhak Rabin. In his letter, Drosnin wrote that according to an obscure code embedded within the Torah, the Hebrew version of the first five books of the Old Testament, Rabin would be assassinated. Guri passed the letter along to Rabin, but alas, no heed was taken, and Rabin was in fact assassinated — a year later.

Convinced that his prediction must have been divinely inspired, Drosnin wrote a book called *The Bible Code* in which he detailed his coding methodology. It was based on the work of Israeli-Latvian mathematician Eliyahu Rips, who had based his work in turn upon that of school teacher Avraham Oren. Others throughout history had dabbled in similar pattern finding, including Isaac Newton. Drosnin's book was a major success, and is about to be followed by its second sequel.

The codes in *The Bible Code* are what's called Equidistant Letter Spacing, or ELS. A very simple example of this is the word *troops*. Take every other letter, and you'll get the word *top*. The word *top* is said to be encoded within the word *troops* as an ELS with a "skip" of 2. That's all there is to it. Now imagine a sentence written out in a grid, without spaces or punctuation, like a giant crossword puzzle with no black squares. A word encoded in this manner will appear in a line, vertically, diagonally, or horizontally, and it can even skip squares. It will often be at an angle something like a knight's move, like 4 squares up and 1 square to the left. With a large enough block of text, it's possible to find just about any word. Add a computer to the mix to do brute force crunching of all the possibilities, and you'll be surprised at how many words crop up. Short words are everywhere. Each time you add a letter to make your target word longer, the number of hits drops dramatically. It's rare to find a word of 7 or more letters.

Now that you know how to find words encoded within a text, what about sentences? Followers of the Bible Code methodology take a pretty liberal approach to this. It is not necessary to find the entire sentence as one long ELS string; that's impossible. All you have to do is find words that appear on the grid near each other, usually near enough to all be viewed on the same grid (but you can make your grid as large as you prefer). Words can go in different directions with different skips. Picture a word search puzzle with a whole bunch of words circled in it, and this is how sentences are found using the Bible Code methodology. There are no rules governing this process, it's completely up to the individual to decide which words to search for within a text, and then place them in the desired order. There are always many other extra words, especially shorter words, scattered through a given grid, so the researcher has plenty of words to choose from to form the desired sentence.

That's how the Bible Code works. If it seems pretty weak and loose, well, you're right, it is. It seemed that way to me, too, so I looked around and purchased a commercial software program that performs these searches. It's called CodeFinder and it cost around $70. There are several available and this was one that looked decent. CodeFinder came with a number of source texts, including the New and Old Testaments, and also *War and Peace.* Really any long text will do, including completely random text. I generated a large file of random text and did some searches on that as well. First I found my name, which is liberally scattered throughout all texts; maybe I'm holy. CodeFinder works in such a way as to store the locations of each word found to facilitate the building of sentences. If you're patient enough, and willing to try all sorts of alternate wordings, and stick with as many short words as possible, you can find just about any sentence you want in any text. Don't get me wrong, it's not quick and easy, you will have to spend significant time building a decent sentence. I played with it long enough to find "I will die on Friday" and "Brian is a cool guy" in the Bible, in *War and Peace,* and in my own random text file. That was enough for me.

The Bible Code proponents have another secret weapon up their sleeve, and that's the Hebrew language. CodeFinder also comes with a copy of the Old Testament in Hebrew. Remember the Indiana Jones movie where he almost steps through the wrong floor tile because of an ambiguity about the spelling of Jehovah? Hebrew has different forms and different spellings of the same word, in some cases a number of different spellings. When Drosnin wrote *The Bible Code,* he took full advantage of these ambiguities to find the maximum number of matches for a given word in constructing his sentences. He has been widely criticized for this. If he'd have stuck to one form of Hebrew or another, many of Drosnin's sentences would be considered misspelled. Even so, to find the name Yitzhak Rabin in Hebrew, he had to use a skip value of 4,772 characters. That covers a massive block of text in which it's possible to find just about any other word.

So, clearly, it is possible to find almost anything you want in almost any text using the ELS method. The nature of entropy means that there will be accidental words and sentences everywhere. When Bible Code proponents find a sentence, how do they know that this sentence was placed there deliberately by some higher power, and was not just another accidental hit? Proving this distinction is really the key to proving that there's any substance to Bible Code claims, and so far, nobody has put forward any reasonable suggestion of what form such proof might take.

Bible Code proponents often point out that Eliyahu Rips co-authored an article about his discovery which was published in a legitimate peer-reviewed mathematical journal, *Statistical Science,* in 1994. What they fail to point out is that this publication was in no way an endorsement of Bible Codes as predictors of future events, let alone divine inspiration. *Statistical Science* is a mathematics journal, it has nothing to do with religion or predictions, and it does not publish research.

The ELS article was published simply as a mathematically challenging word puzzle, a definition under which Bible Code style ELS findings are perfectly legitimate.

But Drosnin says that it's more than that. Much more. Astronomically more. At least, astronomic in terms of its origins. In his first sequel book, *The Bible Code II,* Drosnin states that the Bible was written by — wait for it — aliens. The same aliens, in fact, who brought DNA to Earth and caused life to first develop here. Drosnin believes that the aliens left the key to decoding the Bible Code inside a steel obelisk buried near the Dead Sea, and Drosnin even claims to have gone searching for it himself. Why an elaborately buried key is necessary is unclear, since cheap $70 software off the Internet decodes it just fine.

So what about Drosnin's famous prediction of Yitzhak Rabin's assassination? It does sound impressive, but consider three points. First, it was only one of innumerable predictions that Drosnin made, the rest of which turned out to be nonsense — such as the nuclear destruction of civilization in both 2000 and 2006, and the devastation of Los Angeles by a meteor in 2006. This is the typical tool of the celebrity psychic: Remembering only the hits and ignoring the misses.

Second, at the time that Drosnin made the Rabin prediction, it was a practical certainty that Rabin was going to be assassinated. The hardcore right-wing Jews were as angry with Rabin as were the Palestinians he was trying to make peace with. Pundits said at the time that it was only a question of which anti-peace group was going to get him first. Psychics all over the world predicted his assassination, and Drosnin was lucky enough to be the one who was invited onto the Oprah Winfrey show, even though his prediction provided no useful information about the date or place of the assassination. This PR is the reason that Drosnin is the one whose book became popular, and not some other psychic.

Third and finally, the actual prediction that Drosnin found contained simply the name "Yitzhak Rabin" and a Hebrew word that can mean "assassinate". Shortened, it can also mean "assassin". In a block of text that massive, innumerable shorter words are found, so Drosnin chose "will". Drosnin selected a few words from his palette and arranged them into "assassin will assassinate Yitzhak Rabin", using the word for assassin twice. Note that it could also be arranged into "assassin Rabin will assassinate Yitzhak" or any of numerous other names also found within the block. In short, it's very hard for a critical thinker who understands the ELS coding to conclude that Drosnin found a definitive prediction that Rabin would be killed. Either deliberately or through gross negligence, Drosnin put this foolishness forward as a prediction, and it remains the strongest evidence in favor of the Bible Code.

48. Unconscious Research of Global Consciousness

In this chapter we're going to take a look at a project that has captured imaginations for nearly a decade, the Global Consciousness Project, which posits that events that emotionally affect large numbers of people cause measurable changes in the output of random number generators.

The principal public face of the Global Consciousness Project is Dr. Dean Radin, an electrical engineer and Ph.D. in psychology. Supporters like to say that the project is part of Princeton University, but this is not so. The project director, Roger Nelson, is in the Mechanical and Aerospace Engineering department there, but that's about the whole depth of the connection. Some of Nelson's resources, like the web site, are hosted by Princeton. The project is funded by private donations through the Institute of Noetic Sciences in Petaluma, California.

It is worthy to note that I cannot, in good conscience, criticize Dean Radin. He is said to be an awesome fiddle and banjo player, and the world needs more fiddle and banjo music. So, Dr. Radin, when you read this, know that I am at heart a supporter; and when you put down your random number generator, and pick up your banjo, I'll be in the front row. If you want to do some good in the world, stick with what works. Now let's talk about this Global Consciousness project of yours.

65 people at various locations around the world have a small hardware random number generator, which they call an egg, connected to a computer. All day, every day, each one spits out random numbers, which are regularly transmitted through the Internet to Nelson's server in Princeton, New Jersey. When the researchers choose an event, they pull the data from that time and put it through a series of filters and analyses and find patterns they say are improbably less random. I'm not going to go into all the details of how they do this, it's really boring and

confusing if you're not a statistician, but they do openly publish all their methodology on Nelson's web site at noosphere.princeton.edu. Their theory is that somehow, the collective consciousness of all the emotional or psychological energy of people focused on the chosen event, somehow affects the random number generators. They do not presume to have any hypothesis for how or why this might be possible, or what the mechanism might be, or really any satisfactory answers to any questions that mainstream scientists have asked them. They simply put forth their findings for what they're worth, and they urge you and and I and everyone else to look at their results and hopefully conclude, as they have, that there's something to all of this, and that it's worthy of further research.

The problem is that people outside their lab either fail to reach the same conclusions or find their methodology so flawed that it's pointless to even review the findings. They do publish what they call criticism on their web site, but it's mainly comments and suggestions from their associates. There is not a lot of published criticism of Global Consciousness out there to cite, and one reason is that their theory lacks consistent claims that are specific enough to be tested. Here are two fundamental questions that they must answer and have not:

❖ **What type of event qualifies as "significant"?** They pick events themselves, without any defined criteria. When they choose an event, they fail to test if there are any other simultaneous events in other parts of the world that might override any effect. What happened in Ghana during the OJ Simpson trial? There are no controls over what types of event triggers an examination of the data, and no controls to eliminate prospective events due to conflicting events.
❖ **What type of effect in the data constitutes a result?** Again, no criteria. They maintain no standards for what constitutes a correlation: whether it's a trough or a spike or some other type of anomaly; whether it should happen before, during, or after the event; how long before or after the event it should be found, or what the

duration should be. In fact, their "results" are all over the map.

So, as they look for undefined results from undefined events, they still manage to make additional errors in their methodology. Here are some of the most flagrant:

❖ **The analysis is not blinded in any way.** When something happens, they look at their data until they find patterns. Proper analysis would come from isolated statisticians with no reference indicating a timeline on the data, knowledge of what to look for, or knowledge of what world event is being matched.

❖ **They do not look for alternate causes of their data anomalies.** Sunspots? Cell phone calls?

❖ **They make claims of specific numbers for how they beat chance.** Clearly, it's impossible to have any meaningful metrics, given the lack of standards for scoring or choosing events.

❖ **They make no attempts to falsify their theory.** They should be looking for alternate causes of the anomalies they claim to find in the output from their eggs, such as sunspots or electromagnetic interference from other devices. They should be looking for alternate or additional effects caused by human emotions, like errors in calculators or digital watches. Why not cell phones or toasters? If this effect is real, their eggs would not be the only things affected. Whenever a Global Consciousness event happens, there should be well known and well established failures of, or anomalies in, electric and/or computerized devices worldwide. It's improbable that these supposed effects would seek out and affect only one specific application of common hardware components used in many other devices. They do not look at other species besides humans whose emotions might be responsible for the effects. Why not dolphins or whales, or for that matter ants? Ants comprise the largest percentage of living matter on earth of any group, and ants certainly have collective behavior. If collective consciousness did have a measurable affect on hardware, ants are the first place I would look.

One of their biggest claims to fame is the finding of a massive data anomaly, stronger than any other found, at the time of the 9/11 terrorist attacks, and Radin calculated that it was 6000:1 that this spike in the data was due to chance. Such a finding would make sense if the theory were true (although 9/11 probably didn't bother very many ants). You'll hear this result time and time again if you listen to one of Radin's lectures or read their materials. But you will have to go out on your own to find a dissenting opinion, which can be heard from anyone else who has actually looked at their data. One such person is Jeffrey Scargle of the NASA Ames Research Center, who undertook an analysis on his own time. Scargle's finding on the 9/11 data was "I personally disagree with the conclusion that anomalous effects have been unequivocally established" and "I judge the degree of cogency of all of the results in both (Radin's and Nelson's) papers as low." Scargle attributes their positive findings to the questionable application of an XOR filter to the raw data, the use of a discredited "p-value" test, the lack of blinding, limited choice of likely effects, and a suspicious process that he describes as "data fiddling".

Dr. Edwin May and James Spottiswoode also performed an independent analysis of Radin's 9/11 results. Their conclusion states in part:

> We show that the choice was fortuitous in that had the analysis window been a few minutes shorter or 30 minutes longer, the formal test would not have achieved significance... We differ markedly with regard to the posted conclusions. Using Radin's analysis, we do not find significant evidence that the GCP network's EGG's responded to the New York City attacks in real time. Radin's computation of 6000:1 odds against chance during the events are accounted for by a not-unexpected local deviation that occurred approximately 3 hours before the attacks. We conclude that the network

random number generators produced data consistent with mean chance.

Now let's talk about the elephant in the room. To any reasonable person, the whole concept of global consciousness is ridiculous at face value. This is true of many pseudosciences. But all that should raise is a red flag; people used to think flight was ridiculous too. But when you find red flags everywhere, they start to add up. Let's look back at the chapter *How to Spot Pseudoscience,* and see if there are any other warning signs. Here's one: They make their announcements through mass media, rather than through scientific journals. When respected journals won't touch research, it's a pretty good indicator that there's something wrong. But radio shows like Coast to Coast AM, that promote pseudoscience, are all over it. Another warning is that their claim is based on some unknown form of energy or force. Also, the claim fails the Occam's Razor test. Again, this doesn't prove anything, it's just another red flag. Which is more likely to be true: That there's nothing to the idea of global consciousness, which is what the consensus of mainstream science maintains; or that these few people using tremendously flawed methodology have uncovered something so profound it would change the way we view everything, and is based on some mystical force unknown to science? Another problem is that the claim comes only from one source that's dedicated to supporting that cause. Legitimate research is always successfully repeated by independent labs. When it's not, you have good reason to be skeptical. Global consciousness does pass a few of these tests, but legitimate research and facts always pass all of them.

Now, Dr. Radin, I know I said I wouldn't criticize you, but I do have to take issue with one of your quotes. You said:

There is no kind way to say this, but the most stubborn skeptics do not understand scientific methods or the use of statistical inference, nor do they appreciate the history, philosophy or sociology of science. Their emotional rejection of the evidence seems to be motivated by fundamentalist beliefs of the scientific or religious kind.

This is a classic straw man argument. You're dismissing the rejection of your questionable evidence by calling it emotional and suggesting that it's motivated by a quasi-religious fundamentalist belief in science. OK, whatever. But when you declare that the people who fail to use your methods to find your same results "do not understand scientific methods," you're really pushing credibility. You're not the only person in the world who understands the scientific process. In fact, you don't appear to understand it very well at all. Please, do us all a favor. Foggy Mountain Breakdown. Go.

49. How to Identify a "Good" Scientific Journal

This chapter is about a subject that's as old as debating itself, and it was prompted by the following email from a listener of my podcast:

> *I would love to look a lot of this up on my own, but am unsure about what sources can be trusted. I know you talked about how scientists are not created equal, but as an average person without the background to fully understand the primary sources or the ability to synthesize a consensus without reading meta-analyses, where can I go for reliable information?*

This listener brings up a great point. One guy says "Hey, leprechauns are real, here's an article in a peer reviewed scientific journal that says so," and then someone else replies "No they're not, because there haven't been any such articles in *my* peer reviewed scientific journals." Almost any debate can degrade into "My peer reviewed scientific journal is better than yours."

Now, a really satisfying answer to this question would be "Here, go to www.legitimate-scientific-journals.com, and you can see at a glance if your source is a credible one." Surely there must be some register like that, right? I will dash your hopes with a simple answer: No. There is no such thing as an authoritative list of reputable scientific journals. There can't be. And the reason is that word "authoritative". Who is qualified to be the authority? No one is. No one must be. The moment that any one group is anointed with the ability to declare a source to be legitimate or not, is the moment that we lose objectivity and impartiality.

It is very important to be aware that there is any number of bodies who *do presume to be such an authority*. Approach with extreme skepticism! The only reason anyone would compile such a list is to promote an agenda. Someone once commented on one of my podcast episodes and tried to shoot me down by pointing out that one of my sources was discredited on a web site called sourcewatch.org. Sourcewatch.org, sounds pretty legitimate, sounds like they do good work, sounds like they're out there looking out for our best interests by rubber stamping some sources and discrediting others. But according to critics, Sourcewatch is a two-man operation that endorses only publications following their own narrow political bias. This is a perfect example of what you should expect from any source that attempts to identify itself as a rubber stamping authority. Be skeptical of any group you find whose purpose is to identify reputable journals.

As long as we're throwing around the word *reputable*, I might as well give the somewhat disappointing answer to the listener's question, and tell where you can find a reliable journal. Scientific journals achieve their status only through long histories and good reputations. To be broadly accepted within the mainstream scientific community, a journal must have established a long history of responsible reporting, good quality articles detailing well performed research, and exhaustive peer review. Long standing reputation among the scientists who matter the most. If you're not one of those scientists, it can be difficult to know which research is good, which editors and referees are good, and which journals have a long history of publishing them in good standing. For this reputation to have any meaning, it must stand on its own and not be supported by appearing on some simple list. Unfortunately, *you just have to know;* but I will give you a starting point in a moment.

While it is essential that good journals be peer reviewed, you should be aware that almost every publication hoping for prominence describes itself as peer reviewed. When you hear someone defend their source by stating that it's peer reviewed, be skeptical. By itself, that's meaningless. Think back to our old example of the guy writing a UFO newsletter in his basement who has a couple of his UFOlogist buddies endorse his writing. Suddenly he's "peer reviewed". This is not the type of peer review that carries any meaning within the mainstream scientific community, since the peers have a clear agenda and have not established long histories of scientific acumen by the legitimate community at large. This is an extreme example, but it does accurately portray a lot of what's out there. When you don't know anything else about a journal, the fact that it calls itself peer reviewed cannot, must not, be allowed to carry any weight.

One source that a lot of laypeople are turning to these days is Wikipedia. What about Wikipedia? It's new and it's a very different animal from anything previously available, and is something of a paradigm shift. Wikipedia is not perfect, but it is generally very good. Its principle weakness, so often pointed out by critics, is also its principle strength. Critics of Wikipedia love to charge that any old Joe Blow can go in there and edit any article to say whatever the heck he wants. And this is true, to a point; but they do have multiple layers of redundant checks and balances in place. Every topic has editors, and every edit eventually makes it past several sets of eyeballs. Every article is read by untold zillions of eyeballs, and tempered with suggested edits by many of them. Most of these suggestions are good, and some of them are bad. The volunteer editors include some of the foremost authorities on the subject, and they include crank nitwits, and everyone in between. Wikipedia has tens of thousands of regular editors, over a thousand administrators who enforce the behavior of those editors (eventually weeding out the crank nitwits), and even a judicial committee which resolves any disputes that are not otherwise handled by the process. The underlying software provides a powerful nerve center from which the editors and administrators can track history and changes. This open source process leads to an inevitable effect: Many Wikipedia articles end up being the closest thing to an authoritative consensus that we have on a given subject. Each article continually improves over time until it becomes what Wikipedia describes as the "ideal" article: "balanced, neutral and encyclopedic, containing notable, verifiable knowledge."

When Wikipedia was first conceived, it was a brand new idea that had never been tried on such a scale. No doubt, it had plenty of growing pains. But they've had years to improve the system. They've been dragged through the media more than once over high-profile errors resulting from vandalism, and every day since inception, they've worked to address those loopholes. The process still isn't perfect, but it's one of the most amazing compendiums in human history.

So, I'm going to give our listener a simple answer to his simple question. Start with Wikipedia, or any other encyclopedic resource with a good reputation like Britannica or MSN's Encarta. Nearly always, good articles will include links to additional references (especially in Wikipedia), but these links are of tremendously varying quality. Be careful of their external links, and carefully note why each external reference is being cited. Good articles will often include a section on criticism or skepticism of the subject. Read it.

Note that I'm no doubt going to be criticized for pointing laypeople toward Wikipedia as a starting point for research, mainly due to the usual criticisms of Wikipedia. But, as I said before, Wikipedia's weakness is also its strength, and I do stand by this recommendation, especially for laypeople of a given subject who don't otherwise have the experience to choose a good starting point.

What about identifying which scientific journals are reliable? Since we're not all scientists in the chosen field with the education and experience to know which are the most reputable publications in our field, we need some kind of list. But, as we've discussed, lists are bad things when they come from a source with an agenda. So we turn again to our source with no agenda, Wikipedia. Search Wikipedia for "List of Scientific Journals" and you'll find that they have a page listing a few hundred reputable journals in most scientific fields. Generally, this is an excellent list. The fact that it comes from Wikipedia, and is constantly being revised for objectivity and quality by experts in each field, is its strongest recommendation. Some fields are not listed, and most subsections are partial. You can drill down to find more. But beware: the further you drill down, the broader the quality control becomes, and the more journals of lower repute are included, and more non-scientific fields are listed. If you use this list to gauge the reliability of a source that you see referenced, you are in as good of hands as are available to the inexperienced journeyman; but you must use the list wisely. Stay at the top level, or as close to it as you can. With each click that you drill down, reputation for the listed journals is generally lower.

Again, this recommendation will no doubt be criticized, and the criticism is generally valid. But I maintain that for the average guy off the street, this is the best way to gain a "good enough" grasp of a journal's quality, and to find good research that truly represents the current scientific consensus.

50. THE MIRACLE OF ETHANOL

Once again I'm going to be politically incorrect and point my skeptical eye at something that comes from nature: Ethanol. Ethanol, largely produced from corn in the United States but also able to be produced from a variety of other organic substances, is increasingly being offered as the alternative fuel of choice for drivers.

I'd like to start off by beating my detractors to the punch. Since I'm going to criticize ethanol, I'm going to be called all sorts of names, but mainly I'm going to be accused of being on the payroll of the big oil companies who are afraid of losing business to nature's wonder fuel. So, yes, I'm a corporate stooge, and I'm secretly getting big bucks under the table for doing this. OK? I get it. Save your breath.

There's a tendency when discussing alternative fuels for cars to only look at the tank-to-wheel part of the equation. Tank-to-wheel refers to the part of the fuel cycle involving the burning of the fuel in the engine to drive the wheels. Pump-to-tank refers to the infrastructure needed to deliver the fuel to your car. Well-to-pump refers to the whole process of creating the fuel, regardless of where it comes from, and delivering it to your local gas station. Well-to-wheel is the term that covers the entire process, from the original drilling of the oil to the rubber meeting the road. Whenever you're discussing an alternative fuel, you should always consider all these parts of the process, especially the overall well-to-wheel view. For example, hydrogen is fantastic when you only consider the tank-to-wheel portion. Unfortunately creating the hydrogen in the first place, during the well-to-pump stage, is expensive and generally a net loss of energy; and the infrastructure to deliver the hydrogen to your car in the pump-to-tank stage is non-existent.

Ethanol's major problems come in its well-to-pump phase. The University of Minnesota has concluded that if we converted all the corn we're already growing into ethanol, it would meet only 12% of our gasoline demand. Plus, we're already using the corn we're already growing, so we need to plant more corn to make ethanol. This means more fertilizer, more pesticides, and more International Harvesters. All of those things use fossil fuels, produce waste, and increase greenhouse gases. Growing more corn takes more water, usually in areas where everyone's already fighting over water rights. In Brazil they make ethanol from sugar instead of corn, which makes their equation work better because they have a natural overabundance of sugarcane. Ethanol cannot be transported in pipelines, because even with the best state-of-the-art pipeline technology, there is always water or other contaminants in pipelines and ethanol absorbs water — that's why you can make a scotch & soda. But it's no longer usable as fuel when this happens. Ethanol must be delivered by truck, which is the least energy-efficient way we have to transport liquids, or by rail car. Estimates vary depending upon which lobbying agency you ask, but the well-to-pump stage of ethanol production ranges from 31% efficiency to -200% efficiency. That worst estimate means that you had to burn three gallons of fossil fuel to put one gallon of ethanol into the gas pump, a net loss of two gallons worth of energy.

The best part about ethanol is pump-to-tank. Since it sits in the the same tanks and uses the same pumps at your gas station, there are no changes needed and no added costs.

And now it's time to deal with the biggest elephant in the room: As a fuel, ethanol really sucks. Ethanol's tank-to-wheel performance is abysmal. Its energy content is only about two thirds that of gasoline — 68% of the calorific content, to be exact. If you fill your tank with ethanol, you'll only get two thirds as far as you would with gasoline. To go the same distance, you need to burn more ethanol. Lots more. Let's say you have an average car that gets 25 mpg on gasoline. One day you decide to be environmentally friendly and you fill your tank with E10 (10% ethanol, 90% gasoline) instead. You'll get 24 mpg, a difference which you probably wouldn't notice. But let's say that next week you go down the street to where they sell E85 (85% ethanol, 15% gasoline). That same car is now only going to give you 18 mpg. That's a big drop from 25. Please, if you're going to use E85 on the premise of helping the environment, take the trouble to look up the numbers, and then decide if this is the best way to meet your goal. Don't just trust that because the oil companies make E85 available that it's automatically good. You will need to burn 137% as much E85 to go the same distance as you would on gasoline.

So why is ethanol so popular? Why does the Indy Racing League use it? Is it to reduce our dependence on oil from the Middle East? Since only a small minority of our oil comes from the Middle East, that doesn't seem like it could be the whole reason.

I'd like to relate a short personal story that I think reflects a lot of the pro-ethanol support. A few years ago I went to my 20th high school reunion, and while there I talked with a former classmate whose job was to lobby cities and other fleet operators to switch to ethanol burning buses and cars. By chance I'd just read an article discussing these well-to-wheel ratios, and asked her about it. Before the sentence was halfway out of my mouth, she saw it coming and put up a hand to silence me; and then flew off the handle on a rabid anti-government, anti-American, anti-Western tirade about how capitalism is the cause of all famine and wars, that anyone who earns over $40,000 a year should be taxed over 100%, and that corporations are not defined in the Constitution and are thus illegal. Now, obviously it's a straw man argument for me to bring this up, as nothing she had to say was coherent or even relevant to the topic of ethanol, but I did find it interesting that these were the motivations of at least one professional ethanol lobbyist. I do not believe that she even understood the term well-to-wheel.

But there is more refined support for ethanol out there. Much of the real reason that political candidates are on its bandwagon is economic. Ethanol can be produced more cheaply than gasoline, it's subsidized by the federal government at 51¢ per gallon, and it's exempt from the federal gasoline tax. It makes more financial sense for oil companies to sell ethanol when they can. Ethanol's popularity has little to do with environmental friendliness or improved fuel economy, and more to do with economics and square-state politics. Next time you hear the stumping politician of your choice espousing the production of ethanol, listen to hear if you're being given the whole story, or just another political sound bite.

Bottom line: Keep working on true next generation fuel and power systems. Don't waste time, energy, and money on ethanol.

Brian Dunning is an author, podcaster, speaker, and software investment professional in Southern California. It is all a desperate flurry of activity to avoid anything that smacks of real work.

His free weekly podcast *Skeptoid: Critical Analysis of Pop Phenomena* began in 2006, taking the underdog side of science and rationality amid the overwhelming majority of noise in the media promoting pseudoscience, alternative anything, and unconditional acceptance of the paranormal. The choice between pseudoscience and science is the choice between the developmental stagnation of the Dark Ages, and progress.

Brian takes every opportunity to speak on critical thinking issues at colleges and local groups. In his spare time he enjoys Jeeping in the desert with his family and playing beach volleyball (badly).

7000931R0

Made in the USA
Lexington, KY
23 October 2010